isual Event Detection

THE KLUWER INTERNATIONAL SERIES IN VIDEO COMPUTING

Series Editor

Mubarak Shah, Ph.D.
University of Central Florida
Orlando, USA

Video is a very powerful and rapidly changing medium. The increasing availability of low cost, low power, highly accurate video imagery has resulted in the rapid growth of applications using this data. Video provides multiple temporal constraints, which make it easier to analyze a complex, and coordinated series of events that cannot be understood by just looking at only a single image or a few frames. The effective use of video requires understanding of video processing, video analysis, video synthesis, video retrieval, video compression and other related computing techniques.

The Video Computing book series provides a forum for the dissemination of innovative research results for computer vision, image processing, database and computer graphics researchers, who are interested in different aspects of video.

VISUAL EVENT DETECTION

NIELS HAERING
DiamondBack Vision, Inc
11600 Sunrise Valley Drive
Reston, VA 20191
USA

NIELS DA VITORIA LOBO
School of Electrical Engineering and Computer Science
University of Central Florida
Orlando, FL 32816
USA

Kluwer Academic Publishers
Boston/Dordrecht/London

Distributors for North, Central and South America:
Kluwer Academic Publishers
101 Philip Drive
Assinippi Park
Norwell, Massachusetts 02061 USA
Telephone (781) 871-6600
Fax (781) 681-9045
E-Mail <kluwer@wkap.com>

Distributors for all other countries:
Kluwer Academic Publishers Group
Distribution Centre
Post Office Box 322
3300 AH Dordrecht, THE NETHERLANDS
Telephone 31 78 6392 392
Fax 31 78 6546 474
E-Mail <services@wkap.nl>

 Electronic Services <http://www.wkap.nl>
ISBN 978-1-4419-4907-3

Library of Congress Cataloging-in-Publication Data

A C.I.P. Catalogue record for this book is available
from the Library of Congress.

Printed on acid-free paper.

Printed in the United States of America.

Contents

Series Foreword

Traditionally, scientific fields have defined boundaries, and scientists work on research problems within those boundaries. However, from time to time those boundaries get shifted or blurred to evolve new fields. For instance, the original goal of computer vision was to understand a single image of a scene, by identifying objects, their structure, and spatial arrangements. This has been referred to as *image understanding*. Recently, computer vision has gradually been making the transition away from understanding single images to analyzing image sequences, or *video understanding*. Video understanding deals with understanding of video sequences, e.g., recognition of gestures, activities, facial expressions, etc. The main *shift* in the classic paradigm has been from the recognition of static objects in the scene to motion-based recognition of actions and events. Video understanding has overlapping research problems with other fields, therefore *blurring* the fixed boundaries.

Computer graphics, image processing, and video databases have obvious overlap with computer vision. The main goal of computer graphics is to generate and animate realistic looking images, and videos. Researchers in computer graphics are increasingly employing techniques from computer vision to generate the synthetic imagery. A good example of this is image-based rendering and modeling techniques, in which geometry, appearance, and lighting is derived from real images using computer vision techniques. Here the *shift* is from *synthesis* to *analysis followed by synthesis*. Image processing has always overlapped with computer vision because they both inherently work directly with images. One view is to consider image processing as low-level computer vision, which *processes* images, and video for later analysis by high-level computer vision techniques. Databases have traditionally contained text, and numerical data. However, due to the current availability of video in digital form, more and more databases are containing video as content. Consequently, researchers in databases are increasingly applying computer vision techniques to analyze the video before indexing. This is essentially *analysis followed by indexing*.

Due to the emerging MPEG-4, and MPEG-7 standards, there is a further overlap in research for computer vision, computer graphics, image processing, and databases. In a typical model-based coding for MPEG-4, video is first *analyzed* to estimate local and global motion then the video is *synthesized* using the estimated parameters. Based on the difference between the real video and synthesized video, the model parameters are *updated* and finally *coded* for transmission. This is essentially *analysis followed by synthesis, followed by model update, and followed by coding*. Thus, in order to solve research problems in the context of the MPEG-4 codec, researchers from different video computing fields will need to collaborate. Similarly, MPEG-7 will bring together researchers from databases, and computer vision to specify a standard set of descriptors that can be used to describe various types of multimedia information. Computer vision researchers need to develop techniques to automatically compute those descriptors from video, so that database researchers can use them for indexing.

Due to the overlap of these different areas, it is meaningful to treat *video computing* as one entity, which covers the parts of computer vision, computer graphics, image processing, and databases that are related to video. This international series on *Video Computing* will provide a forum for the dissemination of innovative research results in video computing, and will bring together a community of researchers, who are interested in several different aspects of video.

Mubarak Shah   Orlando
University of Central Florida

Preface

In this book we argue in favor of a bottom-up approach to object recognition and event detection. The underlying principle of the book is that many diverse pieces of evidence are more useful for object recognition and event detection than the most elaborate algorithm working on an impoverished image representation based on, say, edge information. Our approach is motivated by the data processing theorem, which states that the real world possesses a certain amount of information, only part of which we can hope to measure and extract. Processing the extracted information is leaking further information about the world. David Marr's [74] principle of least commitment and Rodney Brooks' [15] subsumption architecture are instances in Computer Vision and Robotics where researchers have consistently applied the data processing theorem.

We present a framework for the detection of visual events from video that is based on three principles, derived from the data processing theorem:

- Extract **rich image descriptions** to provide an expressive internal description of the world.

- **Process the extracted information in a fat, flat hierarchy.**

- **Missing information will prevent crucial inferences,** while additional information disguises relevant information in the worst case.

The structure of this book reflects these principles. We discuss the extraction of diverse image descriptions and their fusion in a fat and flat hierarchy into object recognizing, shot summarizing, and event detecting components. Rich image descriptions based on many diverse sources of information, such as color, spatial texture, and spatio-temporal texture measures greatly simplify object recognition and event detection tasks.

We present object recognition and event detection results for a number of applications and contrast our framework and its components with alternative solutions.

Acknowledgments

Thanks, Wen-Lin for your help, trust, time, and patience; and thank you, Niels Lobo, Mubarak Shah, and Richard Qian for your guidance and support.

Chapter 1

INTRODUCTION

In this chapter we present the principles underlying our approach to visual event detection. We compare how similar problems are approached in other areas of research and how they compare to the approaches taken in field of Computer Vision over the last 30 years. We present the Data Processing Theorem and illustrate its relevance to object recognition, visual event detection, and image and video understanding.

On the one hand the extraction of meaningful information from video appears to be an amazingly complex problem. The often-subtle interactions between objects over time are expressed in terms of a vast number of pixel intensities. To appreciate the variety of abstractions consider the following sample actions and events. While reading them imagine how one might use pixel intensities to detect them: wildlife hunts, kissing, explosions, overtaking on the highway, picking an orange, surfing, entering a building, rock-climbing, ski jumping, a landing aircraft or bird, a rocket launch, platform diving, water flowing or falling, the surf at a beach, a flag in the wind, etc.

Our dilemma is that we must base our inferences on the pixel values we are given, but at the same time we must not get confused by typical variations in the motions, objects, environments associated with each action or event, nor by the intricacies of the compression schemes used to compact the information (e.g., broadcast/cable TV versus video tape, DVD, BetaCam, VHS, YUV-420, etc.)

On the other hand, there is much redundancy that simplifies abstraction tasks significantly. For many actions and events **we do not need complete object or motion descriptions** to establish their occurrence. Landing events, for example, may be detected for many landing objects without knowing their identity. This is probably just as well, since for many actions and events **we may never have complete object or motion descriptions**. For wildlife hunts, for instance, it would be difficult to model the terrain, the predators' shape, motivation, and health, the occlusions, the lighting conditions, the camera parameters, or the effects of the video compression scheme on the content. We might not care about the kind of predator, the predator's limp due to a thorn in its foot, the prey, the prey's speed or global motion parameters, the weather, the camera, or the compression scheme.

Likewise, we may not know or need to know the people who are kissing; we may not know or need to know what exploded; we may not care about the absolute or relative speed and the kind of vehicles that are involved in an overtaking maneuver.

So, how do we tell the relevant from the irrelevant objects and motions, seeing that there are spatial, temporal, and spatio-temporal patterns that are significant, and without which the detection of an event would be very difficult? While we may have a good idea about what it is that defines an event in terms of objects and actions, it may be difficult to specify how to detect the objects and actions. As objects can be engaged in or used for many different actions, many actions can operate on a number of objects.

We will argue throughout this book that all these issues can be addressed in a single, unified approach that

- simplifies the extraction of semantics from video using a flat hierarchy of abstractions,

- facilitates the rapid design and development of intuitive and effective event detectors,

- offers an efficient way to fuse large amounts of spatial, spatio-temporal, motion, and object information.

As a first step, we derive a large number of spatial features from color and texture measures, motion features from the locations of corresponding image regions in multiple frames, and qualitative motion information from spatio-temporal measures. The advantage of such rich intermediate representations is that many objects, actions, and events of interest can be expressed as relatively simple functions of these low-level features. The alternative is a longer chain of modules that extract more specific information based on a smaller set of bottom-up features.

The following sketch of a conventional approach to object recognition is a typical example of this paradigm: *Object recognition requires 2D or 3D model to 2D image alignment/matching/registration, which requires the selection of feature points which in turn is often based on gray-scale based edge detection.*

This approach is plagued by difficult problems:

1 General 2D or 3D model to 2D image alignment is a computationally intractable problem [46]. For n image and n model features there are as many as $n!$ possible ways to match image to model features!

2 Even the general task of selecting candidate feature points from a set of feature points is intractable. There are 2^n ways to chose candidate subsets of feature points from a set containing n features!

3 In practice the above tasks cannot scale to more than a few objects before running into serious computational problems.

4 Note also that these approaches assume that the object(s) to be recognized is(are) actually present somewhere in the image. As a result they will find a best match even if the object of interest is **not** present in the image!

5 Admittedly, there often are constraints that reduce the number of sensible matches between image and model features. On the other hand, edge detection based feature extraction methods often yield large numbers of false positives (features that don't belong to the object), such as shadow edges that have nothing to do with the objects of interest, and large numbers of false negatives (failing to detect many actual features of an object), such as edges, corners, or other feature points that are not visible in gray-scale images, such as color edges, texture boundaries, interest points in, say, an entropy, frequency, orientation, or fractal dimension representation of the image. This increases the number of image features and the size and number of candidate feature subsets to consider, which in turn further complicates image/model alignment and candidate feature selection.

6 Restricting our feature set to edge/corner points has another effect. A certain setting for the parameters of the edge detector causes certain shortcomings in the edge detector's output. Thus the first step in our chain of processing often performs poorly, and all down-stream processes often have to make up for these early mistakes. As long as the crucial information is merely disguised by the edge detector the down-stream processes merely need to figure out how to recover it. When the edge detector completely discards some information it deems meaningless (from *its* point of view) then the down-stream processes have no chance of recovering the information.

Items 1, 2 and 5 above suggest that we gather more information to describe image and model features. This additional information helps us to tell good correspondences from bad ones. Fewer but better correspondences limit the complexity of the search, match, and alignment stages and yield better results. We will be tempted to reduce the redundancy in the vast amount of data contained in images, but this needs to be done very carefully. Once a processing step has eliminated vital information from an image it may take vast amounts of "smarts" to infer that information from other cues. In most cases it is both simpler and more effective not to lose the information in the first place.

Instead of attempting to match all image features to all object features, item 3 suggests that rich image descriptions themselves be used to suggest candidate object hypotheses. Rich descriptions of image regions form signatures that often describe these regions uniquely. These signatures can then be used as keys to classes of objects. If the features we compute are continuous, chances are that the classes they index are similar or related. Kanerva [66] , for example, discusses many amazing facts about high-dimensional spaces . He also shows how a number of imprecise and vague keys (signatures/feature vectors) can be combined to recover the most relevant information from huge databases.

This implicitly addresses item 4. Objects that are not present in an image do not receive any bottom-up support. Therefore, if we are looking for faces there better be support for a face in the image. Without support a bottom-up approach like the one we propose will not produce false alarms. Of course, it still depends on how desperate we are to find an object of interest. Even in a bottom-up approach we can sensitize our thresholds until we find some evidence somewhere. An active vision system can use bottom-up stimuli to find the image area that produces the strongest response for an object class of interest. The active system might want to investigate the corresponding image region for further evidence using knowledge of the geometry or environment of the object of interest. Likewise, it can scour the neighborhood for evidence of objects that would reject the current hypothesis. The important point is that in such a system the bottom-up system does what it is best at, generating object hypotheses, while the top-level controller does what it is best at, logically combining relatively high-level information.

Item 6 suggests that we aim to keep the number of levels in our abstraction pyramid to a minimum. The fewer levels the fewer follow-up changes to higher-level modules. Adding to the level of the architecture laterally (by providing a richer description of the information at this level) either lends further support for existing hypotheses or it helps weed out ambiguous features.

Based on these observations, it is difficult to see why computer vision systems – to this day – ignore these signs and almost exclusively consist of top-down architectures. The factors that motivate top-down approaches might include the following:

- When research into artificial eyes and vision systems started, researchers had to propose specific tasks they would investigate. Once funding became available those specific tasks were then broken down into even more specific modules, with specific input/output expectations. This resulted in a well organized architecture and chain of processing steps that is more concerned with making the data fit the model than with the data indexing the appropriate model.

- Most of the research into computer vision is financed by corporate, or government grants. In return for their investment these organizations often prefer concrete, deterministic, and intuitive results, rather than self-organizing, data derived solutions. Although countless examples existed for over a decade, where neural networks or genetic algorithms out-perform a cumbersome clump of if-then-else logic, it is understandable that anyone who has to make important decisions prefers transparency of the decision making process to "slightly" better performance.

- This highlights a social factor that will likely remain with us for some time to come. We don't trust machines. We'd rather tell them what to do than give them the freedom to experience and act in our world. We have little reason to believe that given the freedom to act freely in our world they would stick to all the "weird" rules we are used to. Admittedly, nobody would have such suspicions about an intelligent artificial vision system, but maybe that is why the time has come to explore a more autonomous bottom-up approach for vision systems.

- Lastly, however we should mention the most likely reason for the top-down approaches that drove computer vision research during the last 30 years. Compute power. It is much cheaper, in terms of CPU cycles and memory, to do lots of logic on a few pieces of data than gathering tens or even hundreds of millions of bytes of information for every image and every frame of a video sequence. Of course, this important factor is changing quickly too.

This discussion motivates the principles of our approach. The underlying principle guiding our work is the data processing theorem, which we will prove in Section 4.

Theorem 1.1 *The real world possesses a certain amount of information, only part of which we can hope to measure and extract. Processing the extracted information is leaking further information about the world.*

This theorem almost states the obvious, and yet if it merely reflects our intuition why don't we approach difficult problems accordingly, more often? Of course, we have no information about things we don't know about. Of course, we learn more about things by sensing or measuring them. Good analysts can convey much of the complexity of some issue in a concise way. But "increasing" information is commonly called "lying" and is not valued at all. Thus, intuitively we already accept the theorem, and yet when we are faced with a difficult problem in computer vision we typically limit the processing to gray-value images, or worse yet, edge representations of gray-value images.

We believe that robust object recognition and event detection is best achieved by extracting rich image descriptions that aim to capture as much information about the world as possible. This strategy is related to Marr's principle of least commitment [74].

Corollary 1.1.1 *Extract rich image descriptions to provide an expressive internal description of the world.*

Our second corollary is related to Occam's razor. Sir Occam suggested around 1275 that the simplest interpretation is the best, and that simple explanations are preferable to equivalent, more complex explanations. The data processing theorem supports this rule, as it states that the information conveyed about the world is a monotonously decreasing function of the number of stages involved in processing the information we gathered about the world. Thus, having extracted rich image descriptions we should avoid long chains of information reducing processing steps.

Corollary 1.1.2 *Process the extracted information in a fat, flat hierarchy.*

A fat, flat hierarchy gives the processing stages maximal access to information through multiple channels. We use fat here to refer to the bandwidth of the available information channels.

Our third corollary states that access to conflicting information is more useful than access to seemingly consistent, but less complete information.

Corollary 1.1.3 *The availability of more complete, albeit conflicting information is preferable to less complete information.*

Again, this corollary relates to Marr's principle of least commitment. It may be *difficult* to detect a relevant signal in a noisy pool of information, but on the other hand, it is *impossible* to "recover" information once it has been eliminated by earlier processing stages. In the worst case a process can always chose to ignore some of the available information in favor of other information. But often down-stream processes can benefit even from confidence information provided by up-stream processes.

Our framework and the structure of this book reflect these three principles. Since event detection heavily depends on the gathering of diverse information, the extraction of rich image descriptions occupies a significant portion in both, our book and our framework. This information is then combined in only three stages to extract object, motion-blob, and event information. All stages except the event detector at the highest level have access to processed and unprocessed information about the frames of the input video. The event detector uses the available object, motion, and shot information to select candidate event models and then chooses the event model that best describes the section of the video.

1.   Rich Image Descriptions

Rich data descriptions are known to simplify classification and recognition tasks, since the potential of simple decision surfaces to separate the classes increases (monotonically) with the dimensionality of the input representation. For example, a problem that is linearly separable with a k-dimensional representation will remain linearly separable if further dimensions are used to describe the problem. Perhaps more interesting is the fact that a problem that is only non-linearly separable in a k-dimensional representation may be linearly separable if further dimensions are added to describe the problem. This is important, since simpler decision surfaces often produce more intuitive classifications, as noted by Sir Occam around 1275: "The simplest explanation is the best."

The following example illustrates this point. Assume we have obtained the measurements shown in Figure 1.1 for daily amounts of rain fall in centimeters and hours of sunlight of, two towns, say A-town, and B-town:

A-town	Precipitation [cm]	sunshine [h]
Monday	9	1
Tuesday	1	8
Wednesday	3	3
Thursday	5	5
Friday	2	2
Saturday	6	3
Sunday	3	7
B-town	Precipitation [cm]	sunshine [h]
Monday	7	5
Tuesday	3	7
Wednesday	6	9
Thursday	9	2
Friday	7	6
Saturday	2	8
Sunday	10	4

Figure 1.1. Precipitation and hours of sunshine in two fictitious towns.

The bar-graph on the left edge of Figure 1.2 shows the precipitation for the two towns. The black bars represent A-town; the white bars represent B-town. The bar-graph shows the significant overlap in precipitation in the two towns. Likewise, the bar-graph at the bottom of Figure 1.2 shows the significant overlap in the hours of sunshine the two towns enjoy. Given just the amount of rainfall or just the number of hours of sunshine we cannot tell the two towns apart, without a large margin of error.

Given both the precipitation and the hours of sunshine we can use a line like the one in Figure 1.2 to tell the two towns apart. The equation of the line is $y = -x + 10$. Values on the line satisfy $10 - x - y = 0$. All measurements for A-town satisfy $10 - x - y < 0$, and they are disjoint from those for B-town which satisfy $10 - x - y > 0$. On average, B-town gets a lot of precipitation

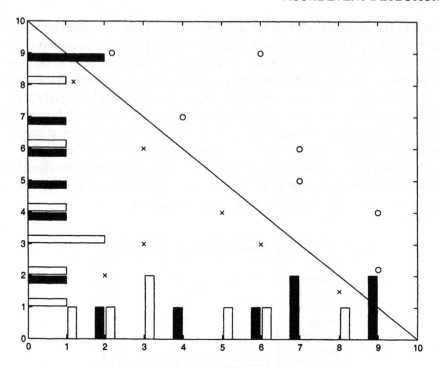

Figure 1.2. The two towns have similar amounts of precipitation, as shown by the bar-graph on the left. The two towns also have similar amounts of sunshine every day, as shown by the bar-graph at the bottom. There is no threshold value or even range of values for either precipitation or sunshine that distinguishes between the two towns. When the two measurements are combined we can use the function described by the diagonal line ($y = -x + 10$) to distinguish between the two towns

and many hours of sunshine, while A-town tends to get less of either. In both towns, though, the two measurements are inversely correlated.

This example indicates an important fact: the use of higher-dimensional representations facilitates better discernibility of patterns and thus simplifies classification. This is an important observation that guides much of what we will discuss in the remainder of the book. Next we will sketch how to use rich image descriptions to extract relevant, meaningful, and versatile intermediate components that facilitate the robust extraction of diverse event and action information .

2.   Constructing Visual Primitives for Object Recognition and Event Detection

Early efforts in computer vision approached image analysis as a language parsing problem [21, 47, 58] with

- syntactic visual measures as the alphabet,

- objects and primitive events as the words, and

- complex events and scene descriptions as the sentences.

While much work has been done at all levels of this language, there are few unifying approaches. Early work focused on the sentences of the language, proposing event recognition and scene descriptions given the locations and identities of objects in the scenes. More recently, the emphasis shifted to the object/word level of the language, proposing object recognition given the locations and types of syntactic primitives. However, throughout the years, there also was a steady effort to discover the most useful syntactic descriptors [52, 53, 76, 108].

The general paradigm was to solve fairly complex tasks in simple environments. We suggest an alternative approach that aims to solve simple tasks in environments of full complexity. This approach requires solutions at all three levels: we need to extract the *relevant* primitives ; construct models for the relevant objects and event primitives from them; and then use the accumulated knowledge to determine more complex events and scene descriptions.

At the lowest level, we need a rich internal representation of visual data. At the intermediate level, the design of simple object and event detectors is based on the internal representation. We found that neural network based non-linear classifiers are well-suited to combine the internal representations to yield robust classification into objects and event primitives. Finally, more complex events and scenes can be described in terms of both the internal representation and the simple object and event detectors.

Extracting higher-dimensional – and hence more powerful – representations of more general use allows us to delay the introduction of task and object dependent constraints in the design of object and event detection solutions. So, let us consider high dimensional representations of the world in more detail.

3. Signature Based Recognition

The senses of smell and taste in animals use a small number of chemoreceptors ("basis functions") to recognize and discern a large number of different odors and tastes. Research in physics, chemistry, and biology has developed automatic methods to analyze and synthesize gases and liquids, describing and creating them as linear combinations of basis molecules (functions). We can automatically *analyze, detect, recognize,* and *discern* the elements and most of the molecules we know, using chromatography, spectroscopy, microscopy, neutron activation analysis, etc.; we can *synthesize* artificial scents and flavors that can fool our senses of smell and taste. Often this can even be achieved by providing only the correct ratios of a few of the most important chemicals (truncated basis sets) of the scent or food to be emulated.

Can we apply these insights from other domains to computer vision ? Can we recognize objects by signatures based on their colors, textures, and motions?

Can we detect events by signatures based on the objects involved and their motions? If so, we have the tools to build a powerful language that lets us describe and detect complex visual actions and events.

We will show in the next chapter that signatures can indeed be used for object recognition, segmentation, and tracking. We will show how a neural network classifier can learn to associate object class labels with feature vectors based on color, spatial, and spatio-temporal frame measures. If the classifier is used to assign object class labels to every pixel of a frame then we obtain meaningful frame segmentations. Finally, tracking objects from frame to frame can be achieved by matching feature vectors representing each candidate object in the current frame against the feature vector representing the object of interest in the previous frame. We obtained good tracking results when each object was represented by an average of the feature vectors at the objects' center. EM-based [32] approaches using much lower-dimensional feature vectors to represent objects have been shown to achieve good tracking results [7, 69].

3.1.   Signatures for Remote Sensing

Researchers in remote sensing [82, 83] have taken a signature based approach to analyze earth surface terrain types. The "signature responses " at a range of frequencies of electromagnetic waves (e.g., ultra violet (UV), visible, infrared, microwave, radar), are used to:

- *classify* terrain into urban areas, saltwater bodies, freshwater bodies, desert, granite soil, limestone soil, shrub, steppe, tundra, deciduous forest, coniferous forest, marsh, swamp, rainforest, etc.

- *detect* the presence of important minerals, metals, chemical warfare agents, etc., and

- *measure* evaporation, algae growth, plankton blooms, soil, water, and air quality, crop conditions, etc.

3.2.   Visual Signatures

While many objects around us may well have signature responses when exposed to a range of electromagnetic frequencies, many practical problems complicate this approach for object recognition.

- For many of the frequencies outside the visual spectrum, no imaging techniques are known. Most ultra violet light is either *absorbed* by matter or passes through largely *unaffected*. Frequencies below IR require *slow* scanning methods and the use of antennae instead of lenses. Their resolution is limited.

- While a relative sensor displacement (e.g., the width of an antenna) is not problematic for terrain scanning purposes obtained via aircraft or satellite at high altitude, the same is generally not true for object recognition tasks.

- The storage and transmission of high-dimensional "signature responses" for each pixel is expensive.

- Large existing visual data bases only show the depicted objects and actions in the visual spectrum; we cannot obtain, say, a UV or infra red representation of an old movie in an archive.

However, it is possible to gather rich object descriptions, even if they can not be obtained by measuring the responses of the objects at a range of frequencies.

Analogous to the remote sensing approach, we can use the "red", "green", and "blue" intensities to represent a pixel. A brightness measure, and opponent color measures, like "red" vs. "green", form further pixel measures, as do higher-order combinations of these pixel representations. The expressive power of these descriptors can be improved further by sacrificing some of the locality of the measures. For example we can augment the descriptions by texture measures based on the area surrounding a pixel and (if available) by qualitative and quantitative motion measures. In this book we present a scalable approach to object and event recognition for computer vision using color, texture, and motion measures to construct a language that can describe image and video data.

4.   The Data Processing Theorem

The motivation behind this strategy – to use rich image descriptions – is formalized in the *Data Processing Theorem*. In short the theorem states that processing data inevitably reduces the amount of information conveyed by it about the world.

Consider the ensemble WDR in which W is the state of the world; D is the data gathered; and R is the processed data, so that these three form a **Markov chain**

$$w \to d \to r.$$

Thus, the joint probability $P(w, d, r)$ can be written as

$$P(w, d, r) = P(w)P(d|w)P(r|d).$$

Then

$$H(W; R) \leq H(W; D)$$

i.e., the information that R conveys about W, $H(W; R)$, is less than the information that D conveys about W, $H(W; D)$. Or in plain English, processed data tells us less about the state of the world than the data we gathered (and in turn, the data we gathered only conveys part of the true complexity of the world). Appendix A shows a formal proof of the *Data Processing Theorem*.

Two consequences follow from this theorem:

- Unless we know that certain data carry **no** information for the task at hand, we should try to preserve as complete a representation of the world as possible; and

- Long sequences of processing steps may reduce information severely (of course, in specific cases it may be the intention to drastically reduce the information, e.g., the steps of Canny's edge-detector: obtain the image → obtain the horizontal and vertical components of the gradient of the image → combine these intermediate results to obtain an edge-strength image → suppress non-maximal gradient points in the resulting gradient image → apply hysteresis tracking → edge image).

5. Object Recognition

An insightful definition of the goal of object recognition was given at the Workshop on 3D Object Representation for Computer Vision [87] in 1994: "... *model-based vision must go beyond pose estimation and into actual object recognition: for model databases containing thousands of objects, we cannot afford to try every model, estimate its pose, then verify its presence in the image using the estimated pose. We must also tackle the difficult problems of extracting the relevant information from images (segmentation), automatically constructing the object models, indexing in sub-linear time the model database and eventually integrating the corresponding modules into working end-to-end recognition systems.*"

We propose the use of color, texture, and motion signatures in a bottom-up evidence pooling approach to alleviate many of the problems mentioned in this excerpt. Signatures can simultaneously index the object identity, and simplify segmentation and tracking. We show how object models can be learned from examples and how a hierarchy of information derived from an image or image sequence can be used to obtain end-to-end event detectors.

The goal of Computer Vision research is to achieve complex tasks in unrestricted environments. Since this was too hard, researchers started off with simple tasks in heavily constrained environments. Segmenting blocks in blocks worlds, given a list of corner locations ([21, 47, 58]), is an example of this early strategy. When a simple task had been mastered, the restrictions on the task were relaxed. Staplers, telephones, and toy-cars replaced the simple blocks [2]. This gave rise to a large number of shape-based representations and alignment algorithms [45]. The computationally intractable alignment algorithms required further simplifications in the environment, often assuming that it contains nothing but the object to be recognized. If eventually a reasonably complex (and usually high-level) task had been achieved, the constraints on the environment were relaxed. This violated the initial assumptions, and the carefully crafted approaches were rendered ineffective. Usually large chunks of the programs had to be reprogrammed. The methods developed for the blocks world example, mentioned above, cannot be used for natural objects or for human made objects that do not have well-defined corners; in fact, even the assumption that corner locations in a blocks world can be obtained robustly and automatically proved difficult.

Approaching vision in this top-down fashion often created chicken-and-egg problems, where a high-level problem depends on good performance of a lower level module, while the low level module cannot function without relevant high-level information. A common chicken-and-egg problem is the dependence of some object recognition "solution" on good segmentations, while the segmentation module cannot produce good segmentations without knowing *what* to segment.

An analogous problem in robotics research prompted Rodney Brooks [15] to propose a subsumption architecture for the construction of robots. Rather than starting off with complex tasks in simple environments, he suggested starting off with simple tasks in environments of full complexity.

The subsumption architecture approach significantly inspires the work described in this book. We propose an end-to-end object recognition approach that uses rich image descriptions as intermediate representations. The image descriptors consist of color and texture measures describing different properties of image regions. While each descriptor in isolation is weak, the combination of a number of them achieves robust object recognition. This approach performs well under a wide range of lighting and imaging conditions, object sizes and orientations, image compression and transformation schemes as well as significant shape variations in the objects. This is important since many non-rigid objects, like sky/clouds, trees, grass, fire, water, rocks, mountains, etc., cannot make use of geometry or shape based recognition schemes. We found this approach to be effective for the classification of natural scenes into categories such as trees, grass, sky/clouds, rock, and animal. We can make use of these classifiers wherever useful, and refine for instance the animal detector where necessary.

A bottom-up approach to object and event recognition reduces the burden on the verification process, by limiting the number of possible models (interpretations) for a given image or video segment. This is an important pre-requisite for a high level knowledge-based search that selects the most relevant model, given both bottom-up evidence and top-down confirmation. In a way the chicken-and-egg problem pair, object recognition and segmentation, are solved together. Anything that looks like a tree, is a tree, or at least part of a tree image region. Anything that looks like a cloud, is a cloud, or at least part of a cloud image region. This approach does not rule out the use of geometric relationships between components of objects to help recognize more complex compound objects. In fact it lends support to the detected components, and thus naturally enables the formulation of plausible hypotheses.

While top-down methods desperately try to find evidence for the known objects (anywhere in the image, at any orientation), bottom-up methods combine evidence and support from the image to index objects and hypotheses.

We argue that the extraction of a rich set of image and video descriptors is the first step toward an extensible, robust, and versatile architecture that can achieve simple tasks in complex environments. The combination of primitives from this internal representation of the world can be used directly to describe and detect much more complex real-world events and actions. As we will show later, it is often sufficient to gather and combine simple color, texture, and motion information to describe complex events, such as animal hunts, rocket launches, etc.

6.   Event Detection

In many ways event recognition is to video data what object recognition is to image data. To describe the content of image data, we need to detect, recognize, and label objects; for video data we need to detect, recognize, and label objects, actions and events, such as   animal hunts [51], entering a room or depositing an object [24], explosions [57], monitoring human behavior in an office environment [6], etc. Although the human visual system can often infer events, like those mentioned above, even from still images, it is generally believed that event detection or recognition from video data is simpler. For example, from a single image of a human made satellite in space we cannot conclude whether it is plunging to earth, heading for Mars, or orbiting earth; a video sequence of the satellite eliminates this uncertainty.

We limit ourselves to detecting **visual events** that consist of visible actions on objects over time, as opposed to recognizing events from still images, or through the extensive use of context. For instance a good photographer might be able to capture a parent's pride at their child's pre-school graduation in a single picture. We cannot hope to have machines make these kinds of inferences any time soon. And, while people enjoy the use of context to interpret events and actions, we better not require that machines develop a sense of "awareness", and "common sense" to "understand" the context before we attempt to detect events and actions. Thus, for the time being, we need to be able to gather visual information about the phases of an event and the objects participating in it. We call such events, *visual events*, and the remainder of this book describes a framework for their detection.

7. Practical Applications

The amount of information in images and video that can be accessed and consumed from peoples' living rooms has been ever increasing. This trend may be further accelerated due to the technological and functional convergence of television receivers and personal computers. To obtain the information of interest, tools are needed to help users extract relevant content and effectively navigate the large amount of information available in images and video.

Existing content-based image and video indexing and retrieval methods may be classified into the following three categories:

1 syntactic structurization,

2 image or video categorization, and

3 extraction of semantics.

For *image retrieval*, work in the first category has concentrated on color, texture, measures of entire images and shape descriptions of selected objects [96, 114]. Work in the second category has focused on categorizing pictorial data into graphics, logos, or images [4]. Work in the third category aimed at classifying images based on their semantic content, e.g., indoor versus outdoor [111], landscape versus cityscape [115], or tree versus non-tree image regions [50].

For *video retrieval*, work in the first category has concentrated on

- shot boundary detection and key frame extraction, [3, 122]

- shot clustering, [120]

- table of content creation [34],

- video summarization [75], and

- video skimming [106].

These methods are computationally simple and their performance is relatively robust. Their results, however, may not necessarily be semantically meaningful or relevant since they do not attempt to model and estimate the semantic content of the video. For consumer oriented applications, semantically irrelevant results may distract the user and lead to frustrating search or browsing experiences. The work in the second category tries to classify video sequences into certain categories such as news, sports, action movies, close-ups, crowds, etc. [63, 116]. These methods provide classification results, which may facilitate users to browse video sequences at a coarse level.

Video content analysis at a finer level is probably needed, to more effectively help users find what they are looking for. In fact, users often express their search items in terms of more exact semantic labels, such as keywords describing objects, actions, and events. The work in the third category has been mostly specific to particular domains. For example, methods have been proposed to detect relevant events in

- football games [60],

- soccer games [121],

- basketball games [101],

- baseball games [67], and

- sites under surveillance [24].

The advantages of these methods include that the detected events are semantically meaningful and usually significant to users. The major disadvantage, however, is that many of these methods are heavily dependent on specific artifacts such as editing patterns in the broadcast programs, which makes them difficult to extend to the detection of other events. A query-by-sketch method has also been proposed recently in [19] to detect motion events. The advantage of this method is that it is domain-independent and therefore may be useful for different applications. For consumer applications, however, sketching needs cumbersome input devices; specifying a query sketch may take undue amounts of time; and learning the sketch conventions may discourage users from using such tools.

In the next chapter we will describe a computational framework and several algorithmic components towards an extensible solution to semantic event detection. The automated event detection framework enables users to effectively find semantically significant events in videos and helps generate semantically meaningful highlights for fast browsing. In contrast to most existing event detection work, the goal of this work is to develop an extensible computational approach for the detection of a range of events in different domains. We will demonstrate the presented approach for the recognition of deciduous trees in still images and for the detection of hunt events in wildlife documentaries as well as landing and rocket launch events in unconstrained commercial videos.

The detection of deciduous trees can be achieved in a two layer architecture. The first layer extracts texture and color measures; the second layer combines the measures to yield image labels indicating the presence or absence of deciduous trees at each image region.

For the detection of hunts in wildlife videos and landing or rocket launches in unconstrained videos, we use a three-level framework. The first level extracts color, texture, and motion features, and detects moving object blobs and shot boundaries. The mid-level employs a neural network to verify whether a moving blob is, or is part of an object of interest, and it generates shot descriptors that combine features from the first level and results of mid-level, domain specific inferences made on the basis of shot features. The shot descriptors are used by a domain-specific inference process at the third level to detect the video segments that contain events of interest.

Therefore, the presented approach can be applied to different domains by adapting only the mid- and high-level inference processes. The low-level feature responses of the video frames form an exhaustive bottom-up representation, and are independent of the task at hand, i.e., the mid- and high-level processing that is based on them.

8. Summary and Overview of the Chapters

In this chapter we presented the principles underlying our approach to visual event detection. We introduced the Data Processing Theorem and emphasized its relevance for object recognition, event detection, and image and video understanding. We looked at some of the tools used by other sciences, such as astronomy, physics, chemistry, and biology, and remote sensing to handle complex non-visual data, and suggested that rich image descriptions form powerful visual signatures that simplify object recognition and event detection. We compared bottom-up and top-down approaches in Computer Vision over the past 30 years.

In Chapter 2 we describe our framework for the design of event detectors and its algorithmic components. Chapter 3 considers alternative classification methods and compares approaches to feature reduction. In Chapter 4, we present experimental results for the detection of deciduous trees in images, the detection of animal hunt events in wildlife documentaries, as well as landing and rocket launch events in unconstrained video tapes. Implementational details are also furnished in Chapter 4. Chapter 2 shows a diagram of our framework for event detection. Throughout Chapters 2 and 4 an icon of the modules of our framework accompanies each section. **The highlighted component of the icon indicates the scope of the discussed issues and results. We realize that the book covers a large number of issues and hope that highlighting the relevant components will offer some useful guidance.** In Chapter 5, we summarize the topics we discussed and consider alternatives and improvements to the components of our framework, before concluding with a discussion on future work in Chapter 5.

Chapter 2

A FRAMEWORK FOR THE DESIGN OF VISUAL EVENT DETECTORS

In this chapter we define visual events, present a framework for the detection of visual events and discuss the framework's components. We describe how we obtain rich image descriptions based on color, texture, and motion measures, and how they are combined in a fat, flat hierarchy to infer object, shot, and event information. Throughout this chapter we will indicate the scope of the discussion by highlighting the relevant component within the framework.

We focus on the classification and detection of non-rigid, amorphous or articulate natural objects, such as animals, trees, grass, sky, clouds, etc., as well as the motions and actions of these objects in natural scenes. Our approach, therefore, has object classification and motion detection components. The object classification component makes use of feature extraction methods based on multi-resolution Gabor filters, the Fourier transform, the Gray-Level Co-occurrence Matrix (GLCM), the fractal dimension, and color. The feature representations of the objects are then classified by a back-propagation neural network. This concludes the task for object recognition in still images. For event detection in video data the classification labels are combined with shot boundary information and frame motion estimates to detect semantic events, such as

1 predators hunting prey,

2 bird, planes, or space shuttles landing, or

3 rocket launches.

In the following sections, we will often use the hunt detection example to illustrate the components of our method.

The problem of detecting semantic events in video, e.g., hunts in wildlife video, can be seen as having three levels as shown in Figure 2.1. At the lowest level we determine the boundaries between shots, estimate the global motion, and express each frame in a color and texture space. We compensate for the estimated global motion between each pair of frames. The earlier frame of each pair is transformed by the motion estimate, and a difference image is produced to highlight areas of high residual error. We assume that this residual error is mostly due to the independent motion of a foreground object. Therefore, the highlighted areas correspond to independently moving objects, which are also referred to as motion-blobs (see Figure 4.15 on page 98).

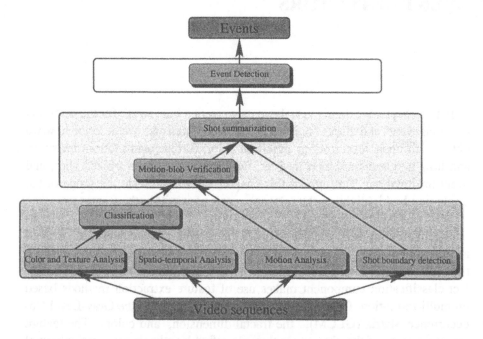

Figure 2.1. The framework of the presented method. For object recognition in still images the method ends after the second level by providing object labels for each image region.

At the intermediate level the detected motion-blobs are then verified with the class labels assigned to that region by a neural network. The neural network uses the color and texture representation of the input obtained by the lower level, and performs a crude classification of image regions. For instance, in the hunt detection example, image regions are classified into sky, grass, tree, rock, and animal regions. We assert that a fast moving animal is being tracked if

1 the motion between two consecutive frames is large,

2 a blob exists that has a high motion residual (motion other than that of the background), and whose motion and position in consecutive frames varies smoothly, and

3 is labeled as an animal region by the network

The intermediate level generates and integrates such frame information and produces a summary for an entire shot. If throughout the shot there was support for a fast moving animal and the location/motion of the animal was found to be stable enough, then the shot summary will indicate that a fast moving animal was tracked throughout the shot.

At the highest level a domain specific analysis of these shot summaries is used to infer the presence or absence of the events of interest. This level will use shot summaries to move between the states of a finite state machine that models the different phases of the event and the conditions on the transitions between the different phases. See, for example, the hunt detection model in Figure 2.15 on page 59.

The small icons of this framework at the beginning of each section indicate the scope of the discussed material, with respect to the framework. The icons should be interpreted as follows:

- The module at the center of the discussion is printed as a bright gray box.

- The video input, at the bottom, and the detected event summaries, at the top of the framework, are printed as dark gray boxes.

- Modules not affected by the discussed material are printed as medium gray boxes.

1. Low-level Descriptors: Color, Spatial Texture, and Spatio-Temporal Texture

As we have argued above, we aim to obtain rich descriptions of the objects in the video, since they form robust and expressive representations of objects under the many lighting, viewing, compression, and within-object variations, that real images and video exhibit. Each pixel is described in terms of color and texture measures. The color measures are the normalized red, green, and blue intensities of the pixel, its gray-value and 2 opponent-color measures, while the texture measures are derived from the Gray-Level Cooccurrence Matrix (GLCM), Fractal Dimension (FD) estimation methods, the Fourier transform, steerable filters, entropy, and a Gabor filter bank. The feature space representation of each *pixel* is classified into a number of categories using a back-propagation neural network. The combination of a rich set of features for each pixel and a back-propagation neural network classifier provide a robust tool that enables the detection of deciduous trees in unconstrained images [50] and the categorization of image regions into classes, such as sky/clouds, grass, tree/shrub, animal, and earth/rock.

An analysis of this rich feature set shows that many features are highly "linearly redundant". Note, however, that we qualified *redundant*. We will show in Chapter 3 why measures of the redundancy and relevance of features or feature sets need to be qualified in this way to be meaningful. For instance, classifiers that are based on the covariance matrix of the features are limited to linear relationships between the features. Other classifiers, such as neural networks, can make use of non-linear relationships between the features, and can therefore often discriminate between classes that appear indistinguishable to classifiers that are based on the covariance matrix. We found that, irrespective of the classifier used, *subsets* of the many features we gather generally yield worse classification results than the entire set of measures. The performance loss due to the omission of a small number of features is almost unnoticeable both subjectively (i.e., as judged by humans) as well as objectively (i.e., as determined by some kind of error measurements). The redundancy in the feature set makes all classifiers more robust to camera, imaging, compression noise, etc. Variations due to these sources of noise often have slightly different effects on seemingly redundant features. Often measures that respond very similarly over a wide range of data differ significantly in a few special cases and thus provide useful discriminatory information to a classifier. For feature sets that utilize less than 90 % of the available features, both subjective as well as objective error measures indicate a deterioration of the classification performance. The neural network described in Section 2 is well suited to combine this set of measures and robustly classify image regions into a number of categories, such as tree, sky, grass, and animal. Except for 6 entropy measures the color and texture features are computed from still frames and motion is included explicitly at a higher-level. The 6 spatio-temporal entropy measures are based on pairs of consecutive frames. An alternative measure of spatio-temporal textures is based on the spatio-temporal auto-regression with moving average (STARMA) model model [110]. The STARMA spatio-temporal model has been successfully used to classify and categorize spatio-temporal textures. The main drawback of this approach is that it is based on large numbers of consecutive frames and requires large amount of memory and processing resources. STARMA based spatio-temporal texture analyses are best suited to stationary spatio-temporal textures that do not change rapidly in the spatial and the temporal domain.

The following subsections will introduce 55 features extracted from 7 fundamental extraction methods based on color (9), the gray-level co-occurrence matrix (28), the fractal dimension (4), Gabor filters (4), the Fourier transform (1), steerable filters (3), and entropy (6). The numbers in parentheses are the number of features extracted from each method.

1.1. Color Measures

Some of the most intuitive measures we can make for each pixel are its intensity and color intensities in the red (R), green (G), and blue (B) bands of a color image. Unfortunately, the intensities of the red, green and blue components are highly correlated and thus are not as useful a representation of a pixel's color, as for instance a hue (H), saturation (S), and intensity (I) decomposition. Sometimes this representation is also called the HSV representation, for hue, saturation, and value. The HSI representation offers more complementary, i.e., less correlated information of the color of the pixel, than the RGB representation. See Section 3. 9 for a discussion of the pros and cons of de-correlating measurements. The Hue component (θ) can be computed by finding the angle between the color of a pixel with respect to the **red** corner of the color triangle in Figure 2.2(a) (see for example [65] for details).

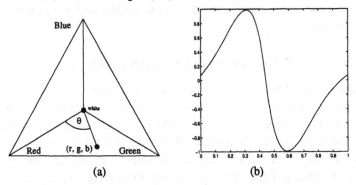

(a) (b)

Figure 2.2. The color triangle and a derived sample function that maps a pixel's hue to its probability of being a leaf pixel.

$$\theta = cos^{-1} \left(\frac{2r - g - b}{2\sqrt{(r-g)^2 + (r-b)(g-b)}} \right)$$

where r, g, and b are the intensity of the red, green and blue components of the corresponding pixel. The **saturation (S)** and **intensity (I)** components are also defined in terms of r, g, and b:

$$S = 1 - \frac{3\,min(r,g,b)}{r+g+b}$$

$$V = \frac{1}{3}(r+g+b)$$

The color-value discontinuity between magenta-red and orange-red (which have maximally different feature values on the hue scale but appear very similar in images) complicates the task unnecessarily. However, using ground truth or otherwise, it is possible to transform the hue values into probability density function. This simple function of the hue may then be used, for instance, to approximate the likelihood of the pixel being a leaf, sky, cloud, or grass pixel (see Figure 2.2(b)).

We also use opponent color measures that contrast the intensities of Red vs. Green ($\frac{Red}{Green+\alpha}$), Red vs. Blue ($\frac{Red}{Blue+\alpha}$), and Green vs. Blue ($\frac{Green}{Blue+\alpha}$), where we used $\alpha = 0.01$ to bound the ratios.

Gray-level images have the same intensity in all three color bands at every pixel and can thus be seen as images with zero saturation and identical hue at every pixel location. Such distortions and transformations are not restricted to the color domain. Many compression schemes for instance deliberately alter both the texture and the color of image regions. In all we use 9 color measures, the gray-level, red, green, and blue intensities, the hue and saturation, and the 3 opponent colors.

1.2.   Gray-level Co-occurrence Matrix Measures

Just because a pixel is bright it need not be part of a sky region in an outdoor image or represent a light in a room. It does, however, say that the corresponding surface in the world either reflects or produces light, which may narrow our search for the right object somewhat. Knowing that the intensity of a pixel is very low is even less meaningful on its own. It could be just about any object in the absence of light. An object recognition system solely based on the intensities of pixels is unlikely to be of much use. Unfortunately, it does not help us much more to know that a pixel represents a green object in the world. It could represent part of a foliage or grass region, a rusty piece of copper wire, a car, or even my favorite mug.

Luckily, we are not constrained to measuring statistics of individual pixels. Texture measures combine a number of pixels and attempt to capture the relationships between their intensities and colors between them. Co-occurrence matrices are 2-D histograms of pixel intensities that indicate how many times pixels of intensity x border pixels of intensity y in each direction. This informal definition of co-occurrence matrices should illustrate their dependence on a definition of neighborhood. In a simple case a pixel might be considered to have only four neighbors, the ones to its immediate north, east, south, and west. To describe a pixel and its neighbors' intensities in this case we would need to specify the intensities of the pixel (i) and that of its neighbor (j), the distance d of that neighbor, and the relative orientation θ of that neighbor, with respect to the pixel. Let $p(i, j, d, \theta) = \frac{P(i,j,d,\theta)}{R(d,\theta)}$ where $P(.)$ is the gray-level co-occurrence matrix of pixels separated by distance d in orientation θ and where $R(.)$ is a normalization constant that causes the entries of $P(.)$ to sum to one.

The northern, eastern, southern, and western neighbors of a pixel would have $\theta = 0^o$, $\theta = 90^o$, $\theta = 180^o$, and $\theta = 270^o$, respectively. The distance d in this case would be 1, indicating that we refer to the immediate neighbors of the pixel. Parameters i, and j indicate the intensities of the two neighboring pixels at this orientation and distance. As an example, consider the gray-level image patch shown in Figure 2.3 and the corresponding co-occurrence matrix

108	107	105	103	107	100
107	100	103	101	105	101
107	106	105	102	108	100
105	106	101	105	107	100
104	105	102	106	107	100
103	104	101	108	107	100

Figure 2.3.   An intensity image patch.

shown in Figure 2.4 for $\theta = 0^o$ and $d = 1$ The bold entry in the table indicates

	100	101	102	103	104	105	106	107	108	109
100	0	0	0	1	0	0	0	0	0	0
101	0	0	0	0	0	2	0	0	1	0
102	0	0	0	0	0	0	1	0	1	0
103	0	1	0	0	1	0	0	1	0	0
104	0	1	0	0	0	1	0	0	0	0
105	0	1	2	1	0	0	1	1	0	0
106	0	1	0	0	0	1	0	1	0	0
107	5	0	0	0	0	1	1	0	0	0
108	1	0	0	0	0	0	0	2	0	0
109	0	0	0	0	0	0	0	0	0	0

Figure 2.4.   The co-occurrence matrix for $\theta = 0$ and $d = 14$ for the image patch in Figure 2.3

that there are two locations in the above image patch where a pixel of intensity 108 has a pixel of intensity 107 as its immediate neighbor ($d = 1$) to its right ($\theta = 0^o$).

While pixel intensity and color band intensities measure first order statistics of a pixel, the gray-level co-occurrence matrix measures second order statistics of a region. In texture classification, the following measures have been defined, see for example [22, 52]:

The **Angular Second Moment (E)** (also called the Energy) assigns larger numbers to textures whose co-occurrence matrix is sparse.

$$E(d, \theta) = \sum_{j=1}^{N_g} \sum_{i=1}^{N_g} [p(i, j, d, \theta)]^2$$

The **Difference Angular Second Moment (DASM)** assigns larger numbers to textures containing only a few gray-level patches.

This and other features use $p_{x-y}(n, d, \theta) = \sum_{j=1}^{N_g} \sum_{\substack{i=1 \\ |i-j|=n}}^{N_g} p(i, j, d, \theta)$

$$DASM(d, \theta) = \sum_{n=0}^{N_g} p_{x-y}(n, d, \theta)^2$$

The **Contrast (Con)** is the moment of inertia around the co-occurrence matrix's main diagonal. It is a measure of the spread of the matrix values and indicates whether pixels vary smoothly in their local neighborhood.

$$Con(d, \theta) = \sum_{n=0}^{N_g-1} n^2 \left[\sum_{\substack{j=1 \\ |i-j|=n}}^{N_g} \sum_{i=1}^{N_g} p(i, j, d, \theta) \right]$$

The **Inverse Difference Moment (IDM)** measures the local homogeneity of a texture. It weighs the contribution of the co-occurrence matrix entries inversely proportional to their distance to the main diagonal.

$$IDM(d, \theta) = \sum_{i=1}^{N_g-1} \sum_{j=1}^{N_g-1} \frac{1}{1-(i-j)^2} p(i, j, d, \theta)$$

The **Mean (M)** is similar to the contrast measure above but weighs the off-diagonal terms linearly with the distance from the main diagonal, rather than quadratically as for the Contrast.

$$M(d, \theta) = \sum_{n=0}^{N_g-1} n \left[\sum_{\substack{j=1 \\ |i-j|=n}}^{N_g} \sum_{i=1}^{N_g} p(i, j, d, \theta) \right]$$

Similarly to the Angular Second Moment the **Entropy (H)** is large for textures that give rise to co-occurrence matrices whose sparse entries have strong support in the image. It is minimal for matrices whose entries are all equally large.

$$H(d, \theta) = - \sum_{j=1}^{N_g} \sum_{i=1}^{N_g} p(i, j, d, \theta) \log \left(p(i, j, d, \theta) \right)$$

Other measures are, **Sum Entropy (SH)**, which uses $p_{x+y}(n, d, \theta) = \sum_{j=1}^{N_g} \sum_{\substack{i=1 \\ |i+j|=n}}^{N_g} p(i, j, d, \theta)$

$$SH(d,\theta) = -\sum_{n=0}^{2N_g-1} p_{x+y}(n,d,\theta) \log\left(p_{x+y}(n,d,\theta)\right)$$

Difference Entropy (DH)

$$DH(d,\theta) = -\sum_{n=0}^{N_g} p_{x-y}(n,d,\theta) \log\left(p_{x-y}(n,d,\theta)\right)$$

Difference Variance (DV)

$$DV = -\sum_{n=2}^{2N_g}(n-DH)^2 p_{x-y}(n,d,\theta)$$

The **Correlation (Cor)** measure is an indication of the linearity of a texture. The degree to which rows and columns resemble each other strongly determines the value of this measure. This and the next two measures use $\mu_x = \sum_i i \sum_j p(i,j,d,\theta)$ and $\mu_y = \sum_j j \sum_i p(i,j,d,\theta)$.

$$Cor(d,\theta) = \frac{\sum_{i=1}^{N_g-1}\sum_{j=1}^{N_g-1} ijp(i,j,d,\theta) - \mu_x * \mu_y}{\sigma^2}$$

Shade (S)

$$S(d,\theta) = \sum_i^{N_g}\sum_j^{N_g}(i+j-\mu_x-\mu_y)^3 p(i,j,d,\theta)$$

Prominence (P)

$$P(d,\theta) = \sum_i^{N_g}\sum_j^{N_g}(i+j-\mu_x-\mu_y)^4 p(i,j,d,\theta)$$

Note that the directionality of a texture can be measured by comparing the values obtained for a number of the above measures as θ is changed. The above measures were computed at $\theta = \{0°, 45°, 90°, \text{and } 135°\}$ using $d = 1$. To make the measures rotation invariant, the average and range over the 4 orientations to obtain 2 features for each type of measure can be used. For further discussion of these gray-level co-occurrence matrix measures, see [22, 52, 117].

1.3. Fourier Transform Measures

The human eye is thought to process some visual signals in the frequency domain [23]. Just like in the spatial domain there are often patterns in the frequency domain that are indicative of one object or another. Therefore, we use Fourier transform based measures to detect patterns in the frequency representation of images. Some measures commonly used with Fourier Based Methods are i) wedge sampling, ii) annular-ring sampling, and iii) parallel-slit sampling. Wedge sampling techniques are frequency invariant measures of the orientation of a texture. Annular-ring measures, on the other hand, are rotation invariant measures of the energy of a texture within a certain band of frequencies. Parallel-slit measures are often used to tile the frequency response of an image or image patch into disjoint regions. Histograms and other statistical measures of the energy within these tiles can often offer orientation and frequency invariant signatures of different textures.

Many textures differ significantly in the domains of the annular-ring and parallel-slit measures; however, for our work we used a method that is often referred to as angular wedge sampling. Fourier transforms (FTs) of images and image patches containing human-made structures often have line or wedge shaped areas of high spectral power that pass through the center as shown in Figure 2.5 (a) and (b). Figure 2.5(a) is an image patch that shows an apartment complex partly occluded by a tree. Figure 2.5(b) shows the corresponding frequency represention for the patch in (a). Summing the power in fixed angular intervals for all directions in the FT of the image lets us separate common from uncommon orientations in the image. Figure 2.5(c) is an image patch that contains nothing but tree foliage. Figure 2.5(d) shows the corresponding frequency representation of the image patch in (c). It is obvious that there are no dominant directions in the image patch. Frequency representations like the one in (d) are uncommon for many human-made objects.

The shaded wedge in Figure 2.6 shows such an angular interval. A circular mask has been imposed so that the power in the diagonal directions is not unfairly biased. Once the power in each angular interval has been determined, we obtain the minimum and maximum angular power and use the normalized ratio $\frac{max-min}{max+min}$ to determine the amount of structure in the patch.

Larger values for this wedge measure indicate greater "regularity" in some direction in the image patch, smaller values indicate less "regularity", in terms of parallel lines, bars and edges. Since we are comparing the ratio between the maximum and minimum value, this measure is rotation invariant. Also, remember that annular wedge sampling methods are inherently frequency invariant. We will show later that this measure, albeit rotation and frequency invariant, is nonetheless a useful measure of the amount of structure in an image or image patch.

Performing the above procedure on fixed-size image patches, we obtain local measures of the regularity of these patches. Very similar results were obtained for patch sizes spanning three orders of magnitude 16×16, 32×32, and 64×64 pixels. This is important, since we have no control over the scale at which the world is depicted.

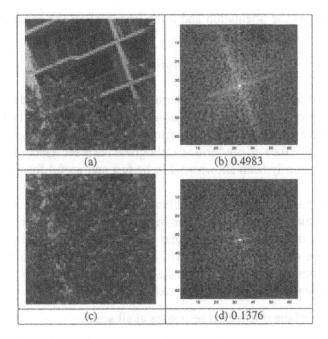

Figure 2.5. An image containing human-made and tree areas (a) and its Fourier Transform (b). An image of leaves of a tree (c) and its Fourier Transform (d). The numbers associated with (b) and (d) are the structure measure (described in Section 2.6) for images (a) and (c).

Figure 2.6. The sum of the power of the Fourier Transform inside the shaded vertical angular interval is a measure of the "structure" present in an image patch.

1.4. Gabor Filter Measures

In the spatial domain the image is described by its 2-D intensity function. The Fourier Transform of an image represents the same image in terms of the coefficients of sine and cosine basis functions at a range of frequencies and orientations. Similarly, the image can be expressed in terms of coefficients of other basis functions. Gabor [42] used a combined representation of space and frequency to express signals in terms of "Gabor" functions or "Gabor" filters:

$$F_{\theta,\nu}(\mathbf{x}) = \sum_{i=1}^{n} a_i(\mathbf{x}) \, g_i(\theta, \nu) \qquad (2.1)$$

where θ represents the orientation and ν the frequency of the complex Gabor function:

$$g_i(\theta, \nu) = e^{i\nu(x\cos(\theta)+y\sin(\theta))}e^{-\frac{x^2+y^2}{\sigma^2}} \tag{2.2}$$

These functions consist of a sinusoidal term with a certain phase and orientation and a Gaussian term that focuses the sinusoidal term to the center of the function. Gabor filters have gained popularity in multi-resolution image analysis [40, 42], despite the fact that they do not form an orthogonal basis set. Gabor filter based wavelets have recently been shown [77] to be fast and useful for the retrieval of image data.

We convolve each image with Gabor filters tuned to four different orientations at 3 different scales. Naturally, there are many ways in which the responses of the filters at the different orientations and scales can be combined to produce useful measures of image textures. We compute the average and range of the four measures at each scale. To make the measurements somewhat scale invariant, we obtain the following four texture measures:

- The average of the orientation responses at all scales.

- The average of the scales' orientation response range.

- The range of the scales' averaged orientation responses.

- The range of the scales' orientation response range.

Convolving an image with a number of rotated versions of a filter kernel can be used to obtain orientation invariant measures. Likewise, convolving an image with a number of scaled versions of a filter kernel can be used to obtain scale invariant measures. As we compute the responses of the measure at the various orientations we are creating samples of a random variable which we may represent by the usual statistics, the mean, median, variance, minimum, maximum, probability density function (pdf), cumulative density function (cdf), etc. We could also lump together a number of these random variables and express their joint behavior (e.g., using their covariance matrix).

Note that the covariance matrix can only measure **linear relationships** between **pairs** of variables. It does not capture the degree or kind of non-linear relationships between pairs of variables and it does not capture any kind of relationship between triples, quadruples, or larger groups of variables. We will come back to this issue in Chapter 3 Section 2.

Although we are interested in the response of the Gabor filter bank to the image in the spatial domain, it is computationally faster to take a "detour" through the frequency domain to achieve our goal. Without loss of generality we assume that the filters of interest are of size $M \times M$ and that the image is of size $N \times N$. Each convolution thus costs us M^2 multiplications and additions.

Since we need to convolve the image with the filter at $N \times N$ locations we end up with a cost of $N^2 M^2$ multiplications and additions per filter. To convolve the image with the entire filter bank of 12 filters we have a cost of $(12)N^2 M^2$ if we take this approach. The size of the filters in our filter bank is $M = 21$. Thus, the cost for convolving in the spatial domain is $(12)(21^2)N^2 = (5292)N^2$ multiplications and additions.

Alternatively, we can construct our filters in the frequency domain (or transform them there) and utilize the fact that convolution in the spatial domain corresponds to multiplication in the frequency domain, and vice versa. The transformation that takes us from the spatial to the frequency domain and back is the Fourier transform. A fast implementation, the Fast Fourier Transform (FFT) has a cost of $O(P \log P)$ multiplications and additions. In our case P is the number of pixels in the image, i.e., $P = N^2$. Substituting this into the FFT function we see that we can transform our image to the frequency domain using $O(N^2 \log N^2) = O(2 N^2 \log N) = O(N^2 \log N)$ multiplications and additions. One of the dualities between the spatial and frequency domain states that convolution in the spatial domain corresponds to multiplication in the frequency domain. Thus, we have a cost of N^2 multiplications per filter. For the 12 filters in the filter bank we end up with $12 N^2$ multiplications. Finally, we need to take the results of the 12 multiplications back to the spatial domain at a cost of $12 O(N^2 \log N)$. Adding all the pieces together, we have $O(N^2 \log N) + 12 N^2 + 12 O(N^2 \log N) = O(N^2 \log N)$ for this approach.

For a typical value for $N = 512$, we have as the cost of convolving in the spatial domain: $(5292) N^2 = (5292)(512^2)$ multiplications and additions, while going through the frequency domain costs us

$$O(N^2 \log N) < c N^2 \log N$$
$$= 9 c 512^2$$

multiplications and additions. In practice $9c$ is much smaller than 5292 which means that the detour through the frequency domain is worth the effort. For the image and filter sizes mentioned we compute the filter responses about 20 times faster via the FFT than in the spatial domain.

1.5. Steerable Filter Measures

In this section we will introduce the concept of steerable filters and demonstrate how they can be used to extract invariant information about image regions. The extraction of invariants from images is central to object recognition. Invariants capture object properties that do not change under certain transformations, such as scale, rotation, translation, or projection. Invariants that are unique to an object or class of objects significantly simplify the object recognition problem of determining which known object is depicted in an image. We will describe a powerful method that can be used to construct steerable filters and we will demonstrate how this method can be used to extract rotation invariant information about image regions. The method we present lets you design your own favorite image filter and it offers an algorithm to determine the response to your filter under many image transformations. We will show how a small number of convolutions with rotated versions of the filter, can be used to determine the response of the filter at **any** angle. Similar results can be obtained to obtain the responses of the filter at **any** scale.

Now, there is no such thing as a free lunch, right? Right. The price we pay for the generality of this method is that we may have to do a little more work than would be necessary for a select special case. The method is based on the fact that any function can be decomposed into an orthogonal set of odd and even basis functions. Any function! Even a 2-D convolution kernel. Of course this little fact comes with its own small print. To say that any function can be decomposed into an orthogonal set of odd and even basis functions, one needs to note that the basis set may need to be infinite. For Computer Vision and image processing applications, however, this does not concern us much. There is no need to approximate convolution kernels to infinite precision when our data – the image – is sampled at a finite resolution.

Since we can approximate any function by a linear combination of basis functions we can convolve the image with the odd and even basis functions, and then use the fact that they are orthogonal to each other and compute the response to the original function of interest.

We will demonstrate how the method can be used to construct a filter that detects step-edges and bars/lines at any orientation.   Since many human-made structures exhibit a large amount of regularity in the form of parallel lines and bars, patches with few dominant orientations are less likely to represent natural objects, such as trees, or clouds, etc. On the other hand, the irregular structure of these natural objects often exhibits a greater variety of weak orientations.

Binning orientations appropriately, we use the *number* and *strength* of different orientations in an image patch to distinguish between patches belonging to human-made scenes (which usually have fewer but stronger distinct orientations) and natural scenes (which tend to have more but weaker distinct orientations). A third rotation invariant measure we gather for each pixel is the ratio of the maximum to the minimum response of the filter to the area around the pixel.

To measure the energy of an image (in terms of line- and step-edges), the convolutions with two filters that are ninety degrees out of phase (i.e., filters that form a quadrature pair) can be squared and summed. To detect lines, one filter type is an even function that can be decomposed solely into cosine terms, while for the detection of step-edges the other is an odd function that can be decomposed solely into sine terms. In the upper row of Figure 2.7, an image of a line is superimposed with three even filters, while in the lower row an image of a step-edge is superimposed with various odd filters. In both rows, the image and – the profile of – the filter are aligned to produce a maximal response.

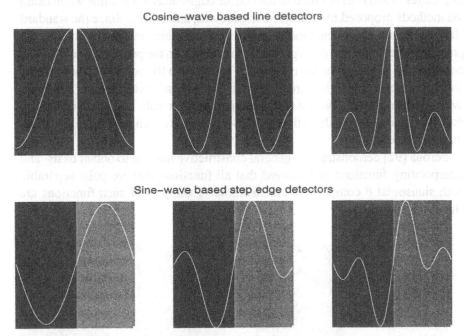

Figure 2.7. Designing line and step-edge detectors by combining cosine or sine terms, respectively. Top row (left to right): $cos(t)$, $cos(t) + cos(2t)$, and $cos(t) + cos(2t) + cos(3t)$ centered on a line. Bottom row (left to right): $sin(t)$, $sin(t) + sin(2t)$, and $sin(t) + sin(2t) + sin(3t)$ centered on a step-edge.

In order to obtain a good orientation resolution of lines and step-edges in the image, one could convolve the image with a large number of orientations of the quadrature pair. However, convolutions are slow, and the accuracy of such an approach depends on a) the sampling frequency and b) the interpolation of the sampled convolutions. Canny showed [17] how the outputs of only two orthogonal filters (the first derivative of a Gaussian in the x and y direction) are sufficient to find the orientation and power of **step**-edges. In order to handle both kinds of edges, Freeman and Adelson [41] use a quadrature pair consisting of the second derivative of a standard Gaussian (the even part of the filter pair) and its Hilbert transform (the odd part of the filter pair). Instead of finding the outputs at all orientations, they introduce the concept of *steering* in which the convolution with a filter at any orientation can be synthesized by a linear combination of the convolutions with a small basis set of filters,

$$F_\theta^{[n]}(\mathbf{x}) = \sum_{i=1}^{n} \sigma_i \, a_i(\mathbf{x}) \, b_i(\theta). \qquad (2.3)$$

Using three basis- and interpolating functions for the Gaussian part and four for the Hilbert transform part of the filter, they showed how to locate both lines and step-edges exactly (as opposed to the double edges detected at a line when using the methods proposed by Canny, Sobel, Prewitt, and others. Since the standard Gaussian has a low orientation selectivity, the kernels are only optimal in the presence of a single line or step-edge in the image. In the presence of more than one edge or line they blur the responses. To improve the orientation selectivity, they suggested using higher-order derivatives of Gaussians as kernels. However, even a fourth order derivative of a Gaussian together with its Hilbert transform does not yield good results if lines or step-edges cross each other at angles other than 90 degrees.

Perona [92] demonstrated a general constructive method to obtain basis- and interpolating functions and showed that all functions that are polar-separable with sinusoidal θ components are steerable. Examples of such functions are shown in Figure 2.8.

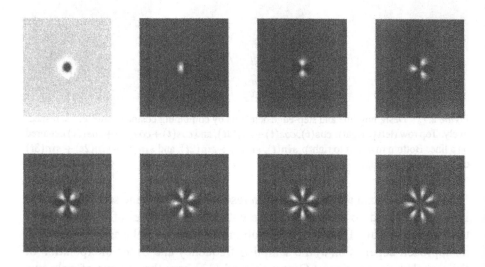

Figure 2.8. Examples of polar separable functions with sinusoidal θ component corresponding to $a_0, \ldots, a_7$.

We used this method to obtain a steerable function set for a quadrature pair (G_{yy}, H_{yy}), where G_{yy} is the second derivative along the y-axis of an elongated Gaussian kernel $G(x, y, \sigma_x, \sigma_y) = e^{-((x/\sigma_x)^2 + (y/\sigma_y)^2)}$ shown on the left of Figure 2.9 and H_{yy} is the Hilbert transform of G_{yy} shown on the right of Figure 2.9.

For multiple occurrences of lines and step-edges at an image location, good angular resolution (orientation selectivity) was obtained when the ratio $\frac{\sigma_x}{\sigma_y}$ was at least $\frac{1}{4}$. Perona [92] shows an efficient method that places the second derivative of the Gaussian in the real part of the complex kernel and its Hilbert transform in the imaginary part.

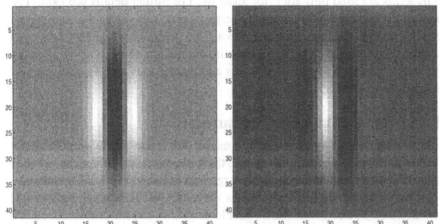

Figure 2.9. Filters used to measure the energy in the image. The second derivative of an elongated Gaussian (left) is used to detect lines in an image. Its Hilbert transform (right) is used to detect step-edges in an image.

The n-term approximation of the function we want to steer can be written as:

$$F_\theta^{[n]}(\mathbf{x}) = \sum_{i=1}^{n} \sigma_i \, a_i(\mathbf{x}) \, b_i(\theta) \quad \forall \theta \in S^1, \ \forall \mathbf{x} \in I\!R^2 \qquad (2.4)$$

where the σ_i weigh the product of the ith filter basis function a_i (the coefficients of the 2D Fourier series) and the corresponding interpolating function b_i (note that the b_i are the frequency basis functions of the Fourier Series).

The values for the σ_i, a_i and b_i are obtained by finding the Fourier series of the function $h(\theta)$, which is the integral of the product of the function with rotated versions of itself:

$$h(\theta) = \int_{I\!R^2} F_\theta(\mathbf{x}) \overline{F_{\theta'=0}(\mathbf{x})} \, d\mathbf{x} \qquad (2.5)$$

where the integral ranges over all 2D space ($I\!R^2$) and $\overline{(.)}$ represents the complex conjugate. Note that $F_{\theta'=0}(x) = F(x)$.

Expanding $h(\theta)$ as a Fourier series we can read off the filter's (2D) basis functions a_i and the corresponding interpolating functions b_i.

$$\sigma_i = \sqrt{h(\nu_i)} \qquad (2.6)$$

$$b_i(\theta) = e^{i\nu\theta} \qquad (2.7)$$

$$a_i(x) \;=\; \sigma_i^{-1} \int_{S^1} \overline{F_\theta(x)} e^{i\nu\theta} dx \tag{2.8}$$

The σ_i terms are used only for error analysis. For further details see [92]. The filters are used to obtain the oriented energy of both step- as well as bar-edges. Although we initially envisaged them to aid the recognition of deciduous trees in winter, when their leaves are missing, the orientation analysis also turned out to be useful for the recognition of leaves and trees in summer, as well as many other natural objects.

1.6.   Fractal Dimension Measures

The underlying assumption for the use of the fractal dimension (FD) for texture classification and segmentation is that images or parts of images are self-similar at some scale. Buildings have large horizontal and vertical boundaries. As we get closer to the building's walls we notice the horizontal and vertical edges of windows and doors. As we get closer yet, we can see the horizontal and vertical edges of the bricks, etc. Cirrus, and Nimbus clouds, coast lines, cities, trees, ferns, etc. all have distinct scales at which they appear self-similar. It is conceivable that distinct object have distinct signatures of the scales at which they are maximally self-similar. Fractal dimension measures are designed to measure self-similarity of objects.

Various methods that estimate the FD of an image have been suggested:

- Fourier-transform based methods [89].

- Box-counting methods [20, 68].

- 2D generalizations of Mandelbrot's methods [88].

The principle of self-similarity may be stated as: If a bounded set A (an object) is composed of N_r non-overlapping copies of a set similar to A, but scaled down by a reduction factor r, then A is self-similar. From this definition, the fractal dimension D is given by

$$D = \frac{\log N_r}{\log r}$$

The FD can be approximated by estimating N_r for various values of r and then determining the slope of the least-squares linear fit of $\frac{\log N_r}{\log \frac{1}{r}}$. We use the differential box-counting method outlined in Chaudhuri, *et al* [20] to achieve this task.

Three features are calculated based on

- The actual image patch $I(i,j)$

- The high-gray-level transform of $I(i, j)$,
$$I_1(i, j) = \begin{cases} I(i, j) - L_1 & I(i, j) > L_1 \\ 0 & otherwise \end{cases}$$

- The low-gray-level transform of $I(i, j)$,
$$I_2(i, j) = \begin{cases} 255 - L_2 & I(i, j) > 255 - L_2 \\ I(i, j) & otherwise \end{cases}$$

where $L_1 = g_{min} + \frac{g_{avg}}{2}$, $L_2 = g_{max} - \frac{g_{avg}}{2}$, and g_{min}, g_{max}, and g_{avg} are the minimum, maximum and average gray-values in the image patch, respectively.

The fourth feature is based on multi-fractals, which are used for self-similar distributions exhibiting non-isotropic and inhomogeneous scaling properties. Non-isotropic scaling refers to scaling that varies with the direction in which it is measured, as opposed to isotropic scaling that has the same properties in all directions. Let k and l be the minimum and maximum gray-level in an image patch centered at position (i, j), let $n_r(i, j) = l - k + 1$, and let $\mathcal{N}_r = \frac{n_r}{N_r}$, then the multi-fractal, D_2 is defined by

$$D_2 = \lim_{r \to 0} \frac{log \sum_{i,j} \mathcal{N}_r^2}{log\, r}$$

A number of different values for r are used and the linear regression of $\frac{log \sum_{i,j} \mathcal{N}_r^2}{log\, r}$ yields an estimate of D_2.

1.7. Entropy Measures

Entropy measures of still images can be used to disambiguate between complex, rough, or messy, on the one hand and smooth, uniform, regions on the other. Natural objects often exhibit less structure than human-made objects both in their spatial appearance as well as their motion. Tree foliage, for instance, is often easily detected in images because it is just about the messiest, most complex, and least structured texture in the real world. Mountains, rocks, earth, grass, meadows, clouds, water bodies, such as oceans, lakes, rivers, streams, etc. all share similar properties. Their surfaces are often rough, irregular and complex. If the surfaces of puddles of water or lakes happen to be calm then they typically reflect the objects around them. A careful analysis of the objects reflected in a calm lake is unlikely to help us detect calm water surfaces. Much higher-level knowledge about the reflected objects is needed to conclude that upside-down clouds, trees, and buildings are not just upside-down clouds, trees, and buildings, but rather are evidence of a reflecting body, such as a calm puddle or lake. Detecting water bodies from the inverted images of the reflected objects is likely to remain a difficult problem for some time to come.

The messiness and lack of structure of many natural objects is not limited to the spatial domain. Over time the appearance of many of these object surfaces changes just as freely as spatially. Even a slight breeze causes leaves and branches to wiggle in the wind, waves to form on a lake, clouds to change their shape, grasses and flowers to sway. Flowing water and oceans never come to rest, and the motions of insects, grazing, gathering, or hunting animals are often governed by too many parameters to be smooth.

The complex appearance of natural objects stands in stark contrast to the rigid and well-structured appearance of many human-made objects. In many cases the walls of buildings are simple uniform surfaces. When they are not then they are often made of regular components, like bricks, and tiles, and have regularly spaced windows. Most edges are either horizontal or vertical, or at least parallel, orthogonal, and co-linear. It usually takes an architect or artist and lots of money to break this regularity and to make buildings appear more "lively".

The rigidity of many human made structures also prevents them from changing freely over time. Cars, machines, bridges, and buildings do not typically deform, sway, wiggle, or bounce in the wind. Thus, human-made objects often differ significantly from natural objects in the spatial and temporal domains. Their spatial and spatio-temporal entropy is a good discriminator between human-made and natural objects.

If V_{max} is the maximum intensity in an image patch, we define its entropy as follows:

$$Entropy = -\sum_{x=0}^{V_{max}} p(x_i) \log p(x_i)$$

where $p(x_i) = \frac{x_i}{N}$ is the ith intensity histogram count n_i divided by the total number of pixels in the image patch (N). For simplicity we will, from now on, drop the index i and the limits of the summation and implicitly assume that all the summations range from 0 to V_{max}. In the following we will also assume that x represents the values of X, and y represents the values of Y.

For video applications we compute 6 entropy measures based on pairs of consecutive frames of video sequences. These 6 measures are

1 The Joint Entropy

$$H(X,Y) = -\sum_x \sum_y p(x)\, p(x,y)\, \log p(x,y)$$

(treating the intensity distributions in the two consecutive frames as two random variables, X (the current frame) and Y (the previous frame)).

2 The Marginal Entropy in the current frame

$$H(X) = -\sum p(x) \log p(x)$$

3 The Marginal Entropy in the previous frame

$$H(Y) = -\sum p(y) \log p(y)$$

4 The Mutual Information

$$H(Y;X) = H(Y) - H(Y|X)$$

between the two frames, where $H(Y|X) = -\sum_x p(x) \sum_y p(y|x) \log p(y|x)$.

5 The Entropy Distance

$$D_H(Y,X) = H(X,Y) - H(Y;X)$$

between the two frames.

6 The Kullback-Leibler divergence measure

$$D_{KL}(X||Y) = \sum_x p(x) \log \frac{p(x)}{p(y)}$$

between the two frames.

Measures 1 through 5 are based on the joint intensity histogram for the gray-values of both consecutive frames, while the Kullback-Leibler divergence is computed treating the gray-values in the two consecutive frames as samples of two populations (the Kullback-Leibler divergence is not a distance since switching the samples may yield different "distance" values, thus violating the associativity constraint distance metrics must satisfy).

For the Kullback-Leibler Divergence (also called the relative entropy) to be computable we must be sure that $p(y) > 0$ whenever $p(x) > 0$, since otherwise the log term becomes infinite and thus reduces the usefulness of the measure. Note that $0 \log 0 \overset{def}{=} 0$. These measures are discussed in much greater detail in [25].

Consider the histogram of gray-values of an image shown on the left of Figure 2.10. The histogram shows the frequency (vertical axis) of each gray-value (increasing from left to right, from 0 to 255) in an image. It can be seen that some of the gray-values between 5 and 30 were not found in the image. If this sample was used, as the Y distribution to compute the Kullback-Leibler divergence then the divergence value would be infinitely negative if the X distribution was non-zero in just one bin between the gray-values 5 and 30. To avoid this problem we use a Gaussian smoothing function for the Parzen windows method to blur the original samples (histograms). The result of the Parzen window method is shown on the right of Figure 2.10. Although not visible due to the discretization in the print the smoothed version of the original

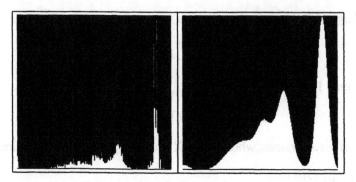

Figure 2.10. The use of Parzen windows to smooth the distribution on the left to enable the stable computation of the Kullback-Leibler divergence.

histogram contains no zero entry bins and the Kullback-Leibler divergence can now be used more reliably to estimate the distance between the distributions X and Y.

An alternative solution would be to replace the Q distribution by a linear combination R of the P and Q distribution, i.e., $R = aP + (1 - a)Q$. This linear combination is zero if, and only if, both P and Q are zero. Note though, that this new formulation measures the "distance" between the distribution P and the distribution R and not the actual distribution Q.

2.   Region Classification

Arbitrating the large number of features gathered at this point is a non-trivial task. Drawing the best conclusions from a pair or triple of features measured for each image pixel is a tough task that may require a solid manual analysis of the data, and may involve many thresholds that need to be adjusted carefully by hand. Trying to combine the 80+ features we compute for each image pixel in any useful manner is a daunting task. The binary decision trees that result from time consuming manual data analyses are typically highly unstable and sensitive to noise. The sequential nature of the decision making process means that a single bad decision somewhere high up in the decision tree will likely prevent us from arriving at the correct conclusions. To avoid the tedious manual analysis that often leads to bad classifiers we make use of a back-propagation neural network to arbitrate between the different features describing the image.   The back-propagation neural network [36] has a single hidden layer and uses the sigmoidal activation function $\Phi(act) = \frac{1}{1+e^{-act}} - 0.5$, where act is the activation of the unit before the activation function is applied. A single hidden layer in a back-propagation neural network has been shown to be sufficient to uniformly approximate any function (mapping) to arbitrary precision [26]. Although this existential proof doesn't state that the best network for some task has a single hidden layer, we found one hidden layer adequate. The architecture of the network is shown in Figure 2.11. The back-propagation algorithm propagates the (input) function values layer by layer, left to right (input to output) and back-propagates the errors layer by layer, right to left (output to input). As the errors are propagated back to the input units, part of each unit's error is being corrected.

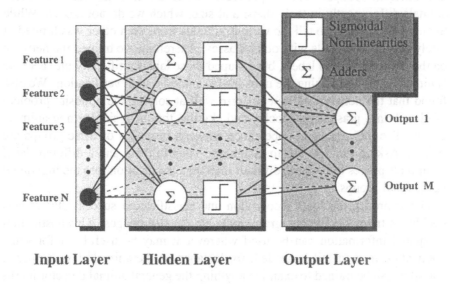

Figure 2.11. The architecture of the back-propagation neural network.

A number of factors prevent error-free results. A few of these complicating factors are that *often there is no correct classification*. For instance, in the case of classifying image regions into tree or non-tree regions, should bushes be labeled as tree or non-tree areas? What if a bush is actually a small tree? In general, it is difficult to label class border pixels correctly, and misclassifications need not all be equally important. Misclassifying a distant herd of animals as trees or rocks is not as severe a mistake as, for example, classifying a nearby lion as sky.

In our tree detection application we trained the network on just two classes, a tree class and a non-tree class. Likewise the landing event detection application uses only sky and non-sky classes. For the rocket launch application we use sky, non-sky, and "exhaust" object classes, while for the hunt application we classify video frames regions into sky/cloud, trees/shrub, grass, animal, and rock regions. For the hunt application we trained the network using a total of 14 labels: 9 animal labels (lion, cheetah, leopard, antelope, impala, zebra, gnu, elephant, and an all-other-animal class) and 5 non-animal labels (rock, sky/clouds, grass, trees, and an all-other-non-animal class) as well as a don't care label that was used to tell the network to ignore border regions between instances of the different groups, which arguably are bad training inputs.

After training, we found that the back-propagation neural network performed well at classifying grass, trees, rocks, sky, and animals each as individual groups. However, it proved difficult for the network to classify lions, cheetahs, leopards, antelopes, impalas, gnus, hyenas, and even zebras, rhinos and elephants each into different groups. This is probably due to the fact that many of those animals differ mostly in their shape and size, which we do not model. While the network's ability to tell the various animals from each other was limited, it rarely confused animals with other classes. Therefore, we trained the network on the different animal labels, but grouped those labels artificially into a single "animal" label when using the network for animal region verification. We also found that the network did not perform well at solving the opposite problem of classifying, grass, trees, rocks, and sky together as a single "non-animal" group. The differences in appearance between instances of these groups are severe. Asking the network to assign one label to them and a different label to animals proves to be more difficult than the classification into the individual non-animal groups.

Our approach to object recognition and classification successfully detects deciduous trees, sky/clouds, grass, rock, etc. in still images. This abstraction of spatial information can be used wherever it may be useful. If for some application more detail is needed, for instance on the animal class, a special classifier can be trained to examine anything the general animal detector marks as animal like. Although the classifier performs only a crude categorization of the image regions it proved to be very helpful for the detection of semantic events in video sequences. This component is useful on its own for object detection and recognition, but we will show next how we can build on this information by integrating it with motion information to extract event information from video.

3.   Global Motion Estimation and Motion-Blob Detection

Now that we have abstracted the spatial information into useful intermediate chunks we will show how this information can be merged with motion information to obtain yet more abstract representations of our data. We will first introduce a few of the most common approaches to motion estimation, before demonstrating that remarkable video event information can be extracted robustly even with very crude motion models. We call image regions, for which we have combined object and motion information, motion-blobs. Until we have gathered enough temporal evidence for these motion-blobs, as we will show in the next section, these motion-blobs are just motion-blobs, i.e., image regions that share spatial and temporal properties.

3.1. Translational Geometry

Translational geometry has two degrees of freedom and can only model translations of image points in the image plane.

$$\begin{bmatrix} x_t \\ y_t \end{bmatrix} = \begin{bmatrix} x_{t-1} \\ y_{t-1} \end{bmatrix} + \begin{bmatrix} Tx_t \\ Ty_t \end{bmatrix}$$

$\begin{bmatrix} x_t \\ y_t \end{bmatrix}$ and $\begin{bmatrix} x_{t-1} \\ y_{t-1} \end{bmatrix}$ represent the location of some image point in the current frame t and the previous frame $t-1$, respectively. Tx_t and Ty_t denote the translation in the x-direction, and the y-direction at time t, respectively. This model is sufficient to model translations of depicted objects or the camera parallel to the image plane, or camera pan and tilt when the camera is infinitely far from the depicted world. In practice, this crude model may also handle camera pan and tilt of cameras mounted somewhat closer than at infinity, but this method will perform poorly when we attempt to estimate translations for anything but a tele-focal camera system.

3.2. Euclidean Geometry

The Euclidean (or rigid body) model has three degrees of freedom. In addition to the camera and object motions modeled by the translational model the Euclidean model also models in-plane rotations.

$$\begin{bmatrix} x_t \\ y_t \end{bmatrix} = R \begin{bmatrix} x_{t-1} \\ y_{t-1} \end{bmatrix} + \begin{bmatrix} Tx_t \\ Ty_t \end{bmatrix}$$

The variables in this model have the same meaning as before, the only difference between the translational transform and the Euclidean transform is that the Euclidean transform can additionally model rotations. Since $R = \begin{bmatrix} cos(\theta) & sin(\theta) \\ -sin(\theta) & cos(\theta) \end{bmatrix}$ only depends on the rotation angle θ, this model has 3 degrees of freedom.

3.3. Similarity Geometry

The similarity transform model has four degrees of freedom. In addition to the camera and object motions modeled by the Euclidean model the similarity transform motion model also models camera zoom and in-plane rotations.

$$\begin{bmatrix} x_t \\ y_t \end{bmatrix} = z_t R \begin{bmatrix} x_{t-1} \\ y_{t-1} \end{bmatrix} + \begin{bmatrix} Tx_t \\ Ty_t \end{bmatrix}$$

The variables in this model have the same meaning as before, except for the new zoom variable z_t, and the new in-plane rotation matrix R, which denote the scale or zoom and the in-plane rotation between the current frame t and the previous frame $t-1$, respectively.

3.4.   Affine Geometry

The affine model has 6 degrees of freedom. In addition to the camera and object motions modeled by the similarity transform model the affine motion model also models in-plane rotations, shear, reflections, etc.

$$\begin{bmatrix} x_t \\ y_t \end{bmatrix} = A \begin{bmatrix} x_{t-1} \\ y_{t-1} \end{bmatrix} + \begin{bmatrix} Tx_t \\ Ty_t \end{bmatrix}$$

The variables in this model have the same meaning as before, the only difference between the similarity transform and the affine transform is that A is not limited to rotation matrices.

3.5.   Projective Geometry

In addition to the camera and object motions modeled by the projective model the projective model also models projections.

$$\begin{bmatrix} x_t \\ y_t \end{bmatrix} = PA \begin{bmatrix} x_{t-1} \\ y_{t-1} \end{bmatrix} + \begin{bmatrix} Tx_t \\ Ty_t \end{bmatrix}$$

The variables in this model have the same meaning as before, except that P is used to model perspective projections.

3.6.   Which Motion Model is Good Enough?

Having briefly discussed the most common motion models, we will show that even a very simple motion model can contribute to robust visual event detection. But first we summarize in Table 2.1 the various transformations and a few of the invariants that can be used to constrain the geometric relationships between objects in the various geometries:

For the examples shown in this book we combined a simple translational model with some scale information.

$$u(x, y) = a_0 + a_2 x$$
$$v(x, y) = a_1 + a_2 y$$

Thus, we estimate translations and zoom but not rotations. There is no particular reason for this choice, other than that we wanted to show that even this simple model suffices to robustly extract visual event information.

The robust recovery of the three parameters has to deal with the following problems:

- Corresponding points in adjacent frames are often far apart (50-60 pixel displacements are not uncommon, peak displacements exceed 100 pixels).

- Interlacing between frames drastically changes the appearance of small objects and textures in adjacent frames.

Table 2.1. A summary of the geometries, corresponding transformations, and associated invariants

	Translational	Euclidean	similarity	affine	projective
Transformations					
translation	X	X	X	X	X
rotation		X	X	X	X
uniform scaling			X	X	X
non-uniform scaling				X	X
shear				X	X
perspective projection					X
Invariants					
global orientation	X				
length	X	X			
angle between lines	X	X	X		
ratio of lengths	X	X	X		
parallelism	X	X	X	X	
incidence	X	X	X	X	X
cross ratio	X	X	X	X	X

- The object and hence the global motion we are trying to estimate is often very large and motion blur eliminates texture in the direction of that motion (of course the motion in this direction is also the motion we are most interested in).

- Objects often need to be tracked under strongly varying lighting conditions and occlusion (e.g., a hunt leading through trees or bushes, a bird flying in and out of the shadow of trees while before landing on its nest, Automatic Gain Control (AGC) of the camera, moving clouds, indoor objects moving past light sources, etc).

- Large foreground objects can severely reduce the areas in which the true background is visible. In an extreme case the true background is entirely occluded by the foreground object(s). In reality it becomes difficult to estimate background motion reliably long before all true background is occluded by the foreground object(s).

- Lens distortion complicates background detection, especially when a wide-angle lens is being used.

- Lack of texture may also complicate motion estimation, since the resulting aperture problems prevent unique motion interpretations.

- Parallax due to camera translation may occlude or reveal previously seen/unseen objects. In such cases matching is difficult to achieve, since the two frames differ by more than two projections of the same scene.

Given the large magnitude of the possible displacements between corresponding patches of adjacent frames an exhaustive search of possible match locations creates unreasonable processing requirements. Therefore, we use a pyramid of reduced resolution representations of each frame, as shown in Figure 2.12. At each level of the 5-level pyramid we consider matches from a 5×5 neighborhood around the location of the patch in the source frame, enabling a maximum matching distance of 62 pixels. At the lowest level of the pyramid,

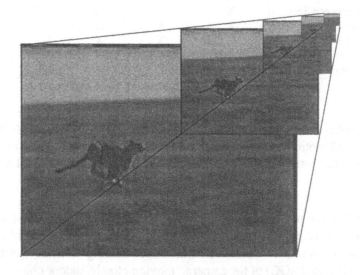

Figure 2.12.   A five level pyramid representation of a video frame.

i.e., the full resolution representation of the frame, the patches used for matching are of size 64×64. Patches from uniform areas often result in erroneous displacement estimates. To avoid matching such patches, patches with insufficient "texture" are discarded. We use a 2D variance measure to determine the "amount of texture".

$$var_x = \sum_{y=0}^{n}(\sum_{x=0}^{m}(p(x,y) - p(.,y))^2 - q_x)^2$$

$$var_y = \sum_{x=0}^{m}(\sum_{y=0}^{n}(p(x,y) - p(x,.))^2 - q_y)^2$$

where p is an $m \times n$ image patch, $p(x, .)$ and $p(., y)$ are the means of the xth column and y^{th} row of p, and q_x and q_y are the means of $((p(x, y) - p(x, .))^2$ and $(p(x, y) - p(., y))^2$ for all x and y within p, respectively. We compute motion estimates at each of the four corners of a frame, as shown in Figure 4.14(a) in Chapter 4. Bad motion estimates are often due to matching errors made high up in the pyramid that are subsequently not recovered by the lower levels. Since the motion of the tracked objects does not vary drastically between consecutive frames (i.e., their (apparent) acceleration is small) we also use the previous best motion estimate to predict the location of the four patches in the next frame. A limited search in a 5×5 neighborhood around this predicted location, improves the motion estimates in many cases. Therefore, we obtain eight motion estimates, one pyramid-based estimate for each of the four patch locations, and one for each of the four estimates based on a limited search around the predicted match locations. Since some patches may not pass the "texture" test, described above, there may be fewer than eight "valid" motion estimates. For a pair of frames we consider only the "valid" motion estimates and determine the "correct" motion estimate to be that, which corresponds to the motion that yields the lowest overall error. That is, for each "valid" motion estimate we determine how well a match it represents at each of the four patch locations. The motion estimate corresponding to the smallest sum of errors at all four patch locations is taken to represent the "correct" motion estimate for this pair of frames.

The normalized dot product between a source patch $P1$ and matched patch $P2$ is equal to the cosine of the angle (α) between the two patches (vectors) $P1$, and $P2$:

$$cos(\alpha)_{P1,P2} = \frac{\sum_{i,j} P1(i,j) P2(i,j)}{\sqrt{\sum_{i,j} P1(i,j)^2} \sqrt{\sum_{i,j} P2(i,j)^2}}$$

With respect to the three particular tasks of detecting hunts in wildlife documentaries and landings or rocket launches in unconstrained video we would like to point out the following:

- Most of the footage is taken with a tele-focal-lens at a great distance to the objects of interest. For our motion analysis, we therefore assume an orthographic camera model, in which the camera pan and tilt appear as plain translations. However, we found that the assumption of uniform background motion is also acceptable for other applications. Applications for which this approach causes problems might have to conduct a more refined motion analysis (e.g., [103, 62]).

- Motion estimates based on the *feature space representation* of the frames are very similar to those obtained on the original *color* frames. A more advanced approach might re-estimate the motion after the moving object has been identified and removed from the frames. It might also be beneficial to use the classification information to select and group features in the frames to enable better motion analysis (e.g., by locating possible depth discontinuities or motion and object boundaries), or better depth/distance/shape recovery (e.g., by providing suitable feature points for which correspondences can be established reliably, given the higher-level knowledge).

- Although the described motion estimation scheme is sufficient for our purpose a multiple hypotheses Kalman filter based approach, as described by [5, 37] might yield more consistent results.

- Alternative camera motion estimation schemes like Video Tomography methods [1] achieve similar motion estimation results. However, since these methods are based on projections of entire frames they are sensitive to large moving objects.

- The most frequent motion estimate in a histogram of a larger number of motion estimates obtained for many image regions may make the background motion estimation more robust.

- Methods that estimate motion and simultaneously segment the video into foreground and background regions can construct background mosaics. Using these mosaics segmentation and motion estimation can be simplified by estimating frame-to-mosaic motion instead of frame-to-frame motion, (see [103, 62]).

The earlier frame of each pair is transformed by the motion estimate, and a difference image is produced to highlight areas of high residual error. The difference image $D(x, y)$ is obtained by summing the "red", "green", and "blue" components at corresponding locations in both frames:

$$
\begin{aligned}
D(x,y) \quad = \quad & F_r^1(x,y) + F_g^1(x,y) + F_b^1(x,y) \\
& - F_r^2(x,y) - F_g^2(x,y) - F_b^2(x,y)
\end{aligned}
$$

Here, $F_r^{(1)}(x,y)$, and $F_g^{(2)}(x,y)$ are the "red" component of frame 1 and the "green" component of frame 2, respectively. The lack of color in a frame will reduce this approach to finding intensity differences between the two frames. We assume that this residual error is mostly due to independent object motion of an object. Therefore, the highlighted areas correspond to independently moving objects which are also referred to as motion-blobs (see Chapter 4 Figure 4.15). Areas with low residual error are assumed to have motion values similar to those of the background and are ignored. The independent motion of objects on the other hand usually causes high residual errors between the current frame and the following motion compensated frame. Therefore, we can make use of a robust estimation technique to obtain an estimate of the object location within the frame. This estimation technique iteratively refines the mean x and y values dependent on the residual error within a fixed size neighborhood around the mean values for the entire difference image. The robust estimation method was developed in [97] for real-time human face tracking. Here, we describe how more general objects, other than human faces, are handled. Based on the frame difference result, the algorithm constructs two 1D histograms by projecting the frame difference map along its x and y direction, respectively. The histograms, therefore, represent the spatial distributions of the motion pixels along the corresponding axes. Figure 2.13(a) illustrates an ideal frame difference map where there is only one textured elliptical moving object in the input sequence, and the corresponding projection histograms.

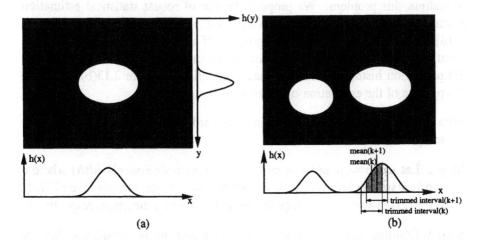

Figure 2.13. (a) Two 1D histograms constructed by projecting the frame difference map along the x and y direction, respectively. (b) Robust mean estimation for locating the center position of a *dominant* moving object.

The instantaneous center position and size of an object in the image can be estimated based on statistical measurements derived from the two 1D projection histograms. For example, a simple method to estimate the center position and size of a dominant moving object in an input sequence is to use the sample means and standard deviations of the distributions. More specifically, let $h_x(i), i = 0, 1, \ldots,$ and $h_y(i), i = 0, 1, \ldots,$ denote the elements in the projection histograms along the x and y direction, respectively. Then the object center position (x_c, y_c) and object width and height (w, h) may be estimated as:

$$x_c = \frac{\sum_i x_i h_x(i)}{\sum_i h_x(i)},$$

$$y_c = \frac{\sum_i y_i h_y(i)}{\sum_i h_y(i)},$$

$$w = \alpha \left[\frac{\sum_i (x_i - \mu_x)^2 h_x(i)}{\sum_i h_x(i)} \right]^{\frac{1}{2}},$$

$$h = \beta \left[\frac{\sum_i (y_i - \mu_y)^2 h_y(i)}{\sum_i h_y(i)} \right]^{\frac{1}{2}}$$

where α and β are constant scaling factors.

However, the object center position and size derived from the sample means and standard deviations may be biased in the cases where other moving objects appear in the scene. It is therefore necessary to develop a more robust procedure to address this problem. We propose the use of robust statistical estimation routines to achieve robust measurements for object center position and size [118]. More specifically, the center position of a dominant moving object in an input sequence is estimated based on the robust (trimmed) means of the two 1D projection histograms in the x and y directions. Figure 2.13(b) illustrates the process of the estimation of the motion center.

Step 1 Compute sample mean μ and standard deviation σ based on all the samples of the distribution.

Step 2 Let $\mu_t(0) = \mu$ and $\delta = max(a\,\sigma, b * sampleSpaceWidth)$ where a and b are scaling factors, e.g., $a = 1.0$ and $b = 0.2$, and $sampleSpaceWidth$ is the image-width and image-height in the x and y direction, respectively.

Step 3 Compute trimmed mean $\mu_t(k + 1)$ based on the samples within the interval $[\mu_t(k) - \delta, \mu_t(k) + \delta]$.

Step 4 Repeat Step 3 until $|\mu_t(k + 1) - \mu_t(k)| < \epsilon$ where ϵ is the tolerance, e.g., $\epsilon = 1.0$. Denote the converged mean as μ^*.

Step 5 Let center-position $= \mu^*$.

In addition to the robust estimation of the object center position, we propose the following routine for robust estimation of object size. The method first re-projects the frame difference result in a neighborhood of the located center. It then derives the object size based on the robust (trimmed) standard deviation. Given the robust mean μ^* and δ obtained from the above center locating routine, the routine for estimating the size in either x or y direction is as follows.

Step 1 Construct a clipped projection histogram H^{clip} by projecting the color filtering map within the range $[\mu^*_{opp} - \Delta, \mu^*_{opp} + \Delta]$ in the opposite direction, where μ^*_{opp} is the robust mean in the opposite direction and Δ determines the number of samples used in the calculation.

Step 2 Based on H^{clip}, compute the trimmed standard deviation δ_t based on the samples within the interval $[\mu^* - \delta, \mu^* + \delta]$.

Step 3 IF $H^{clip}(\mu^* + d\delta_t) \geq g \; H^{clip}(\mu^*)$ OR $H^{clip}(\mu^* - d\delta_t) \geq g \; H^{clip}(\mu^*)$, where e.g., $d = 1.0$ and $g = 0.4$, THEN increase δ_t until the condition is no longer true.

Step 4 Let $size = c \; \delta_t$ where c is a scaling factor, e.g., $c = 2.0$.

Before we show how this motion information is used to detect motion-blobs we briefly summarize our motion estimation approach.

- We only attempt to estimate approximate motion information. We will show in the remainder of the book that this is sufficient to construct robust visual event detectors.

- We assume that we only need to gather motion information for the dominant motion-blob. If motion information needs to be obtained for more than one object we can repeat the above process after the dominant motion-blob has been masked from each video frame.

- We have shown how fast and robust statistical methods can be used to detect the dominant motion-blob in video data.

- We do not need to construct precise camera, object, or background models to detect visual events.

- We use Frame-to-frame differencing between consecutive frames to detect independently moving foreground regions. Background differencing based methods can produce better foreground/background segmentation than the segmentations resulting from frame-to-frame differencing [103, 62]. However, we have shown that simple but robust methods can often compensate for the inferior segmentation results and still allow the construction of robust event detectors.

4. Motion-Blob Detection and Verification

The motion-blob information (spatial, spatio-temporal, and motion information) computed by the motion analysis module just described in Section 3 is combined with the region-classification results of the classifier from Section 2 to determine the identity of the motion-blob object. For further processing of the video sequence we assert that the object dominating the area of the highest residual motion error, as determined by the classifier, is moving with the motion determined by the motion analyzer.

During shots, which show no foreground objects the motion-cued motion-blob detector locks on to random spurious noise in each frame. motion-blobs whose trajectory is too eratic to be a legitimate object are ignored. This filter is based on the assumption that real objects cannot move too far between consecutive frames of a video sequence. Real objects have a mass and thus their inertia will prevent sudden motion reversals and excessive acceleration. A similar argument can be made for the camera. Erratic motions are therefore a good indication of the lack of tracked foreground objects.

If on the other hand there are multiple candidates for foreground objects we mostly rely on the assumption that the dominant motion-blob is of greatest interest. We could also make use of the rich object descriptions available to us to disambiguate between competing frame regions. Say, we were just tracking a lion, i.e., a yellow object of a certain (noisy) size, shape, location, and texture. It is unlikely that all those characteristics change suddenly from one frame to the next and the tracked object turns in into something that has significantly different color, size, shape, location, and texture.

In fact tracking the motion-blob from frame to frame can be achieved by matching feature vectors representing each candidate motion-blob in the current frame against the feature vector representing the motion-blob of interest in the previous frame. We obtained good tracking results when each motion-blob was represented by an average of the feature vectors at its center. EM-based [32] approaches using much lower-dimensional feature vectors to represent objects have been shown to achieve good tracking results [7, 69].

However, we did not have to make explicit use of the extensive information we have gathered about the objects in the video, to achieve robust event detection results. Better use of the gathered spatial, spatio-temporal, motion, and region-classification information accompanying each motion-blob demonstrates one of the untapped resources the presented approach possesses.

5. Shot Detection

While shot boundaries may tell us little about the semantic content of the individual shots they separate, they can help us to interpret and limit multi-shot events. To detect shot boundaries we examine consecutive video frames. Each video frame is converted to a Hue, Saturation, and Intensity (HSI) color format, and represented by a 24 bin histogram with 16 bins for the hue, 4 bins for the saturation, and 4 bins for the intensity, as shown in Figure 2.14. The intersection between any 2 consecutive histograms is computed, normalized and compared to a threshold. If the histograms of the two frames are identical the intersection will be perfect and the returned intersection value

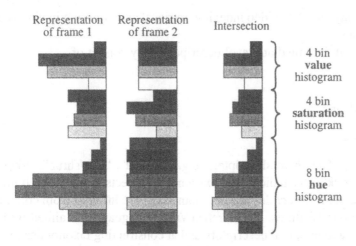

Representation of frame 1   Representation of frame 2   Intersection

4 bin **value** histogram

4 bin **saturation** histogram

8 bin **hue** histogram

Figure 2.14. Shot boundary detection via histogram intersection.

will be 1.0. If on the other hand the two histograms are maximally different, i.e., for each pair of corresponding bins at least one is zero, then the intersection will be minimal and the returned value will be 0.0. We found that using this scheme a threshold of 0.75 achieved good shot detection performance. A more elaborate algorithm might consider frames other than the most recent frames to avoid gradual fades to black, fades to white, cross-fades, dissolves, and other shot transitions. The most obvious choices are a fixed reference frame, or a window of frames across which histogram differences are computed.

6. Shot Summarization and Intermediate-Level Descriptors

We are using a simple color histogram based technique to decompose video sequences into shots. To avoid missing important events in extended shots, we also *force* a shot summary every 200 frames. Finally a third kind of shot boundary is inserted whenever the direction of the global motion changes. Shot boundaries of this last kind ensure that the motion within shots is homogeneous, as far as the direction of the motion is concerned. Each shot is then summarized in terms of intermediate-level descriptors. The purpose of generating intermediate-level shot summaries is two-fold. First, the shot summaries provide a way to encapsulate the low-level feature and motion analysis details so that the high-level event inference module may be developed independent of those details, rendering it robust against implementational changes. Second, the shot summaries abstract the low-level analysis results so that they can be read and interpreted more easily by humans. This simplifies the algorithm development process and aids the stepwise abstraction of content and thus may facilitate video indexing, retrieval and browsing in video database applications.

In general, intermediate-level descriptors may consist of

- *object*,

- *spatial*, and

- *temporal*

descriptors. The *object* descriptors, e.g., "animal", "tree/shrub", "sky/cloud", "grass", "rock", etc. indicate the existence of objects in the video frames. Remember that the *object* descriptors themselves are inferred from color, spatial, spatio-temporal features. Research in Visual Languages has aimed at formalizing spatial and temporal descriptors and at constructing taxonomies of objects. The *spatial* descriptors represent the location and size information about objects and the spatial relations between them in terms of spatial prepositions, such as, "inside", "next to", "below", etc. [31, 33]. The *temporal* descriptors represent motion information about objects and the temporal relations between them in terms of temporal prepositions such as "while", "before", "after", etc. [31, 33]. Much work has also been done on combining spatial and temporal descriptors into action description [81, 85, 113], such as the motion of an agent "towards" another, the "crossing" or "approaching" of two agents, or the "dropping", "lifting", "lowering", "pushing" of an object by an active agent, etc. Most of these are classical Artificial Intelligence approaches and proceed in a top-down manner starting from the English language [81] or taxonomies of motions, actions, objects, or relationships between objects [31, 33, 85, 113].

The shot summary consists of two parts, the general Shot Information part shows general statistics extracted for this shot. The second part is domain specific. For wildlife hunts the Hunt Information part consists of inferences based on the statistics for the hunt detection application.

The first row of the general Shot Information part of the summary shows whether the shot boundary corresponding to this shot summary was real, i.e., whether it was detected by the shot boundary detector, or if it was forced because the maximum number of frames per shot was reached or the global motion has changed. The next two rows show the first and last frame numbers of this shot. The following measurements are shot statistics, i.e., the average global motion over the entire shot on row four, and the average object motion within the shot on row five. The next four rows measure the initial position and size, as well as the final position and size of the detected dominant motion-blob. The third last row shows the smoothness of global motion where values near 1 indicate smooth motion and values near 0 indicate unstable motion estimation.

```
------------ Shot Information -------------
Forced/real shot summary                       : 0
First frame of shot                            : 64
Last frame of shot                             : 263
Global motion estimate (x,y)                   : (-4.48, 0.01)
Within frame animal motion estimate (x,y)      : (-0.17, 0.23)
Initial position (x,y)                         : (175,157)
Final position (x,y)                           : (147,176)
Initial size (w,h)                             : ( 92, 67)
Final size (w,h)                               : (100, 67)
Motion smoothness throughout shot (x,y)        : ( 0.83, 0.75)
Precision throughout shot                      : ( 0.84)
Recall throughout shot                         : ( 0.16)
```

The detection of a reversal of the global motion direction, described above, is based on a long term average of the motion estimates around the current frame, indicating a *qualitative* change in the global motion. The smoothness measure described here, on the other hand, provides a *quantitative* measure of the smoothness of the estimated motion. Finally, the last two rows show the precision and recall averaged over the entire shot. As defined in Section 6, the precision is the ratio of the number of animal labels within the detected dominant motion-blob versus the size of the blob, while the recall is the ratio of the animal labels within the detected dominant motion-blob versus the number of animal labels in the entire frame.

The Hunt Information part of the shot summary shows a number of predicates that were inferred from the statistics in part one. The shot summary shown above summarizes the first hunt shot following a Forced shot boundary. The summary is indicating that it is Tracking a Fast moving Animal and hence, that this could be the Beginning of a hunt. The Tracking predicate is true when the motion smoothness measure is greater than a prescribed value and the motion-blob detection algorithm detects a dominant motion-blob. The Fast predicate is set to true if the translational components of the estimated global motion are sufficiently large in magnitude,

```
------- Hunt Information -------
Tracking                           : 1
Fast                               : 1
Animal                             : 1
Beginning of hunt                  : 1
Number of hunt shot candidates :     1
End of hunt                        : 0
Valid hunt                         : 0
```

and the Animal predicate is true if the precision, i.e., the number of animal labels within the tracked region, is sufficiently large. (The recall measure discussed earlier should also be a useful cue, but we have no data to support or reject our intuition.) The remaining predicates are determined and used by the inference module as described below.

For the landing event detector the event specific descriptors indicate that Tracking is engaged, that the horizontal motion is Fast, that the motion-blob is Descending, that the motion-blob is a non-sky Object in the image, that there is Sky below the object, that the control is in the Approach, the Touch-down, or the Deceleration stage.

```
----- Landing Information ------
Forced/real shot summary        : 1
Tracking                        : 0
Fast horizontal motion          : 1
Descending                      : 1
Object                          : 1
Sky below object                : 1
Approach                        : 0
Landing                         : 0
Deceleration                    : 0
First frame of shot             : 41
Last frame of shot              : 80
```

For the rocket launch event detector both the shot information as well as the event specific descriptors needed adjusting. The Shot Information part of the summary contains the same information as for the other events plus information on the amount of sky, clouds, exhaust, and ground in the frame, above the center, and below the center.

```
------------ Shot Information --------------
Camera motion estimate (x,y)                    : ( 0.00, 0.00)
Within frame object motion estimate (x,y)       : ( 0.00, 0.11)
Initial position (x,y)                          : ( 82,126)
Final position (x,y)                            : ( 78,154)
Initial size (w,h)                              : ( 88, 68)
Final size (w,h)                                : ( 84, 67)
Smoothness (x,y) of motion throughout shot      : ( 1.00, 0.95)
Precision throughout shot                       : ( 0.87)
Recall throughout shot                          : ( 0.21)
Amount of sky in frame                          : (0.41593)
Amount of sky above center                      : (0.36687)
Amount of sky below center                      : (0.27255)
Amount of clouds in frame                       : (0.00000)
```

```
Amount of clouds above center          : (0.00000)
Amount of clouds below center          : (0.00000)
Amount of exhaust in frame             : (0.00000)
Amount of exhaust below center         : (0.00000)
Amount of ground in frame              : (0.32117)
```

For the rocket launch the event specific part indicates that Tracking is engaged, that there is significant Horizontal motion or Vertical motion, that the motion-blob is Ascending, that the motion-blob is a non-sky/clouds Object, that Sky, Clouds, Exhaust, or Ground are visible, that there is Sky/Clouds above/below the center of the frame, that Exhaust is visible below the center of the frame, that Clouds/Exhaust are appearing, that the Ground is disappearing, and whether the control is in the Ignition, the Lift-off, or the Flight stage. See Section 4.3 for an example and further explanation.

```
--------------- Launch Info ----------------
Shot type                        : Frame time-out
Tracking                         : 1
Horizontal motion                : 0
Vertical motion                  : 0
Ascending                        : 0
Object                           : 1
Sky                              : 1
Clouds                           : 0
Exhaust                          : 0
Sky or clouds above center       : 1
Sky or clouds below center       : 1
Exhaust below center             : 0
Clouds appearing                 : 0
Exhaust appearing                : 0
Ground visible                   : 1
Ground disappearing              : 0
Ignition                         : 0
Just saw ignition                : 0
Lift-off                         : 0
Flight                           : 0
Flight candidate                 : 0
Just saw flight                  : 0
First frame of shot              : 1
Last frame of shot               : 40
```

7.   Event Inference

Having described the low and intermediate-level abstractions we can now describe spatio-temporal events in terms of these image abstractions in an intuitive and straight forward manner.   This is demonstrated with the following three examples, the detection of hunts in wildlife documentaries, and the detection of landings and rocket launches in unconstrained video. The events are detected by an event inference module which utilizes domain-specific knowledge and operates at the shot level based on the generated shot summaries. The rules for the inference module are deduced from actual footage of commonly available commercial video. The first event, hunts in wildlife documentaries, depends on the extraction of object and motion characteristics, in roughly equal proportions. Landing events share few object properties but are rich in motion characteristics. Rocket launches on the other hand are rich in object characteristics while the extraction of reliable motion information is difficult.

7.1.   Hunt Events

The rules used for the detection of hunts in wildlife documentaries reflect the fact that a hunt usually consists of a number of shots exhibiting smooth but fast animal motion which are followed by subsequent shots with slower or no animal motion. In other words, the event inference module looks for a prescribed number of shots in which

- there is at least one animal of interest,

- the animal is moving in a consistently fast manner for an extended period, and

- the animal stops or slows down drastically after the fast motion.

Figure 2.15 shows and describes a state diagram of our hunt detection inference model. Although this seems to be a rather crude description of the characteristics of hunts, it allows robust detection of hunts in wildlife documentaries as we will see in Chapter 4 Section 4.4. The success of these simple rules lies in the fact that animals in nature do not run around for no reason. If they do run then it is either because they are being chased or because they are chasing other animals. Humans easily recognize hunts in wildlife documentaries even from severely degraded video data, such as reduced resolution, compression and blur. In many cases it is impossible to infer that the tracked, fast-moving object is a predator, prey or indeed an animal. Occlusion by tall grass often hides their legs, and often the animal occupies only a few tens of pixels in the frames of the video sequence. It is unlikely that from such degraded representations of the animals involved in a hunt it is possible to extract meaningful shape information about

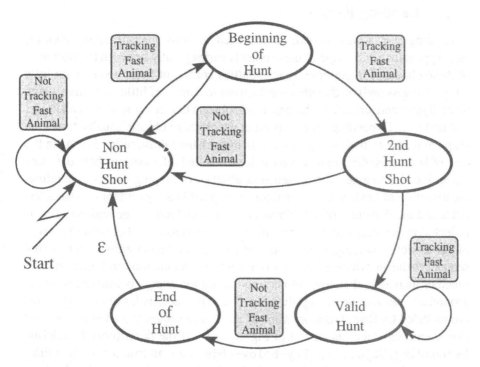

Figure 2.15. The state diagram of our hunt detection method. Initially the control is in the Non-Hunt state on the left. When a fast moving animal is detected the control moves to the Beginning of Hunt state at the top of the diagram. When three consecutive shots are found to track fast moving animals then the Valid Hunt flag is set. The first shot afterwards that does not track a fast moving animal takes the control to the End of Hunt state, before again returning to the Non-Hunt state.

the non-rigid, articulate objects and their motions in unknown environments and camera motions. All these restrictions and limitations of the characteristics of hunts in commercial wildlife videos restrict any hunt detection approach, to little more than crude object and motion characteristics.

Automatic detection of the properties and sequences of actions in the state diagram is non-trivial and the low-level feature and motion analysis described earlier are necessary to realize the inference. Since many events can be defined by the occurrence of the involved objects and the specification of their spatio-temporal relationship, the presented mechanism, of combining low-level visual analysis and high-level domain-specific rules, are applicable to detect other events in different domains. In Section 4.4, we provide an example and further explanation for using this inference model for hunt detection.

7.2.   Landing Events

Landing events may involve objects such as birds, aircraft, space shuttles, etc. Appearance and shape of these objects varies greatly between the instances of these classes, for example, space shuttles consist of large bodies with little wings, tail-less owls on the other hand appear to consist of little more than wings when flying, and aircraft are human-made objects that occur in almost all colors and textures. Depending on the speed at which these objects usually travel the shape and size of their body and wings may change drastically. Just as in the case of hunts it is often unnecessary to have detailed information about the exact object identity to assert the presence or absence of landing events. Therefore, the detection of landing events is likely to depend heavily on motion character-istics and the detection of sky/cloud and non-sky/cloud image regions. This is reflected in the stages and the conditions on the transitions between the stages of the model of landing events. In broad terms the model aims to detect shot sequences during which a formerly flying non-sky/cloud motion-blob first turns much of its potential energy into horizontal motion energy before touching the ground and slowing down significantly. These characteristics of landing events are modeled by four stages, an Approach, Touch-down, Deceleration, and Non-landing stage, as shown in Figure 2.16. If the descriptors Tracking, Descending, Object, and Sky-below-object are all true for the first time, we assert that the current shot could be the Approach phase of a landing. When the control is in the Approach state, the Tracking, Descending, and Object descriptors are true and the object has a Fast-horizontal-motion component, the control moves to the Touch-down state. From this state the control moves to the accepting Deceleration state when the Tracking and Object flags remain set but neither the Fast-horizontal-motion nor the Sky-below-object flags are set. A sequence of shots that does not contain at least an Approach, Landing, and Deceleration phase is not considered a landing event. The landing event ends after the first shot in the Deceleration phase.

7.3.   Rocket Launches

Rocket launches are another example of an event that is best described with-out detailed object and motion characteristics, but that depends heavily on the detection of certain key objects and key motions. If we had to describe the visual characteristics of rocket launches to a two or three year old child it is difficult to see how we could avoid mentioning the rocket engines' exhaust, human-made objects such as the rocket and the launch pad, clouds, and sky. Furthermore it is difficult to define the shape or appearance of launch pads or rockets. Some rockets consist of a single tube shaped object, while the space shuttle has two rocket boosters, a large Hydrogen tank and the shuttle itself.

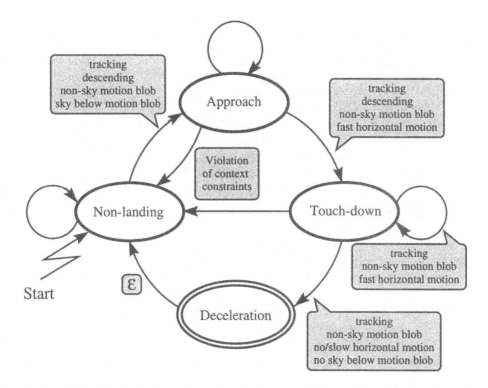

Figure 2.16. The state diagram of our landing detection method. Initially the control is in the Non-Landing state on the left. When a descending object, surrounded by sky, is tracked the control moves to the Approach state at the top of the diagram. When a descending object is tracked and found to be moving with a fast horizontal motion component, the control moves to the Touch-down state on the right of the diagram. Tracking a slow moving or stationary object that is not surrounded by sky causes the control to move the Deceleration state at the bottom before returning to the Non-landing state.

Older or future rockets as well as rockets of other countries may have yet other shapes and appearance. So again it seems best to use a coarse model that captures the salient characteristics of rocket launches without the distractions of a more detailed model. In particular it proves difficult to extract motion information reliably. After the ignition of the rocket engines large amounts of clouds may be created, which depending on the relative camera viewpoint may occlude the rocket itself. Their non-rigid deformation and expansion severely complicates the detection of background/foreground motion.

Our rocket launch event module has four states, Ignition, Lift-off, Flight, and Non-launch, as shown in Figure 2.17. If the descriptors Sky-visible, Ground-visible, Clouds-forming, are all true while there is no motion other than that of cloud regions in the video frames then control moves to the Ignition state. When the control is in the Ignition state, and the

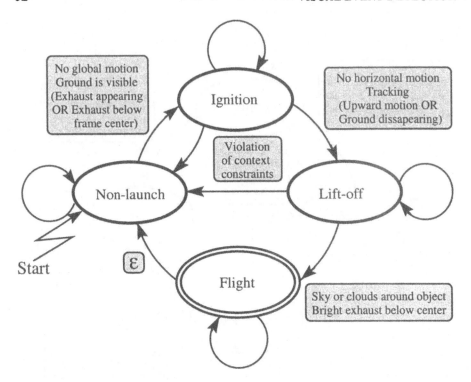

Figure 2.17.   The state diagram of our rocket launch detection method. Initially the control is in the Non-launch state on the left. If sky and ground are visible at the top and bottom of the video frames, respectively, and the only motion is due to developing cloud regions then the control moves to the Ignition state. When a non-sky motion-blob can be tracked in front of a sky background with an upward motion and no horizontal motion then the control moves on to the Lift-off state. Finally, if the tracked non-sky motion-blob continues it's (mostly) upward motion, the ground disappears from the frames and a bright exhaust plume can be seen then the control moves to the accepting Flight state, thus concluding the rocket-launch.

Tracking, Object, Sky-visible and Upward-motion descriptors are true while the horizontal-motion is not set, the control moves to the Lift-off state. From this state the control moves to the accepting Flight state when the Tracking, Object, Sky-visible, and Upward-motion flags remain set but the Exhaust-below-object flag appears and the Ground-visible flag disappears. A sequence of shots that does not contain at least an Ignition, Lift-off, and Flight phase is not considered a rocket launch event. The launch event ends after the first shot in the Flight phase.

8. Summary of the Framework

In this chapter we defined visual events to be events that involve objects, whose actions are visible in video sequences. We presented a framework for the detection of visual events that consists of three components:

1 At the lowest level we extract rich image descriptions in terms of color, texture, motion, and shot boundary information

2 At the intermediat level we combine color, spatial texture and spatio-temporal texture to classify image regions, and we combine motion information with the inferred object information to obtain motion-blob statistics. The motion and motion-blob statistics are gathered and combined with the shot boundary information to generate shot summaries

3 At the highest level these shot summaries are used to navigate a finite state machine based event model, that determines whether there is sufficient evidence to assert that a certain event has been detecte

We discussed the framework's components in detail. In particular we described how we obtain rich image descriptions based on color, texture, and motion measures, and how they are combined in a fat, flat hierarchy to infer object, shot, and event information.

Chapter 3

FEATURES AND CLASSIFICATION METHODS

In this chapter, we discuss alternative classification methods, ways to determine the relevance of features, and feature de-correlation. Since the feature extraction stage is the slowest of the modules, we consider classification without preprocessing in the first part of this section. Next we consider linear, quadratic, and eigen-analysis techniques for the determination of good subsets of features and classification. we will argue that the best method to select subsets of features is based on the same non-linear method as that used for classification. There is little benefit to use one approach to decide on the importance of the features and another approach to classify images. Some systematic approaches are presented which show that the best solutions involve the consideration of an exponential number of combinations of features. The results of the various classification methods are shown in Chapter 4.

1. Classification without Preprocessing

Since finding the feature space representation is costly we trained a Convolutional Neural Network (CNN) [72] on raw images for classification without preprocessing. CNNs are used to apply the same set of weights to each patch of an image, as shown in Figure 3.1. Hence the name Convolutional Neural Network. The motivation for using CNNs is that in image processing there is little reason to operate in one way on one part of the image and in a different way on another. In general there is little importance to the spatial location of objects in the image, since the camera could always be moved to depict the objects in a different part of the image. Applying the same neural network to each patch of an image thus reduces the number of weights/parameters that need to be trained to achieve the desired input/output mapping. The results of the this method are shown in Chapter 4 Section 4.4 on page 99.

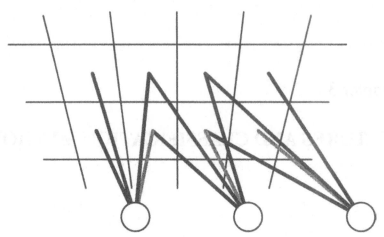

Figure 3.1. This figure shows part of an image grid and three locations of a convolutional neural network classifier. At each of these locations the same weights (shown here in different gray-tones) are multiplied with different pixels from the image. Since the weights remain the same irrespective of the location in the image, as is the case when convolving the image with a filter, this approach is called the convolutional neural network classifier.

2.   Linear Relationships between Pairs of variables

Covariance and correlation matrices only measure *linear* relationships between *pairs* of variables. Therefore, methods based on them neither capture non-linear relationships between pairs of variables (see left of Figure 3.2) nor can they be used to evaluate the linear or non-linear relationships between more than two variables (see right of Figure 3.2). Covariance-based classifiers, such

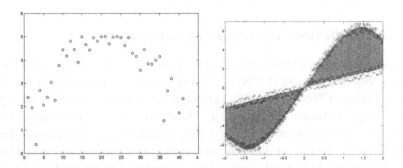

Figure 3.2. The plot on the left shows a sample where X and Y have zero covariance and yet are dependent. The plot on the right shows two different data samples (one with a linear distribution and one with an "S"-shaped one). Both are equally linearly correlated. But while one data set has a plain linear relationship between its two variables, the second actually has a cubic relationship between its two variables. The shaded area indicates the area where the two distributions differ.

as linear, quadratic, or eigen-analysis based classifiers, cannot tell these two data sets apart. Non-linear classifiers on the other hand can exploit the differences in the actual distributions and tell them apart.

3. Linear Analysis

If we restrict ourselves to linear classification methods, Fisher's Linear Discriminating Functions (LDF) [38], Maximally Discriminating Functions [79], or the Best Linear Discriminating Function [100] can be used. To determine the linear redundancy of features we can use the F-test or Wilks test [98]. Fisher's Linear Discriminating function is widely used to classify an observation into one of two classes. An observation is taken to belong to class 1 if

$$(\bar{x}_1 - \bar{x}_2)^T S_{pl}^{-1} \left(x - \frac{1}{2}(\bar{x}_1 + \bar{x}_2)\right) > \ln\left(\frac{\hat{\pi}_2}{\hat{\pi}_1}\right)$$

and to belong to class 2 otherwise. Here $\bar{x}_i$ denotes the mean of class i, S is the pooled sample covariance matrix, x is the observation to be classified, and π_i reflects the prior information we have about the likelihood of the observation belonging to class i.

Linear classifiers are inferior in power to non-linear classifiers, and both the linear classifiers as well as the tests for redundancy require the calculation of the inverse of the class or pooled covariance matrices. Since our set of features is very *linearly* redundant, the associated covariance matrices are too singular to enable the stable computation of their inverse. Section 1 in Chapter 4 shows the performance of this classifier for training and test images.

4. Quadratic Analysis

According to the quadratic discriminating function an observation is taken to belong to class 1 if

$$x^T(S_2^{-1} - S_1^{-1})x - 2x(S_2^{-1}\bar{x}_2 - S_1^{-1}\bar{x}_1) + (\bar{x}_2^T S_2^{-1}\bar{x}_2 - \bar{x}_1 S_1^{-1}\bar{x}_1) >$$

$$\ln\left(\frac{|S_2|}{|S_1|}\right) + 2\ln\left(\frac{\hat{\pi}_2}{\hat{\pi}_1}\right)$$

and to belong to class 2 otherwise. Here $\bar{x}_i$ denotes the mean of class i, S is the pooled sample covariance matrix, x is the observation to be classified, and π_i reflects the prior information we have about the likelihood of the observation belonging to class i.

While quadratic discriminating functions [107] generally yield better results than linear discriminating functions, they too require the calculation of the inverse of the covariance matrix of the features used. But since the covariance matrix of the highly *linearly* redundant feature set cannot be inverted we cannot use quadratic discriminating functions either. Section 1 in Chapter 4 shows the performance of this classifier for training and test images.

5.   Eigen-analyses

Eigen-analyses can be used to express a data set in terms of an orthogonal basis set, called the eigen-vectors of the data set. Assume a video shot consists of k frames of size $M \times N$, and that we represent each of the k frames by the intensities of its pixels, strung up into a vector of length $(M)(N) \times 1$. Now, let D be a matrix whose k columns are the k frame vectors, and let C be the correlation matrix $C = DD^T$. The eigen-vectors and eigen-values of C are then defined as

$$Cx_i = \lambda_i x_i$$

where x_i is the ith eigen-vector and λ_i is the eigen-value corresponding to the i_{th} eigen-vector. Since D has dimensions $(M)(N) \times k$ C is $(M)(N) \times (M)(N)$. C has $(M)(N)$ eigen-vectors and eigen-values. The eigen-vectors are all orthogonal to each other, and thus each frame in D can be expressed as a linear combination of the eigen-vectors. The eigen-analysis produces orthogonal eigen-vectors under the constraint that none of the vectors have length zero, i.e., not all their elements are zero. The eigen-values accompanying each eigen-vector indicate their importance in the reconstruction of the data in D. For instance, if C is, as we stated above, the correlation matrix DD^T, then the first (most significant) eigen-vector will often contain the mean data value for each variable. In our example each variable represents a pixel in a frame, while the k frames are the k observations we made. In this case the first eigen-vector will contain the mean intensity at each pixel location averaged over the entire shot.

But this need not always be the case. If instead of computing the correlation matrix C we compute the covariance matrix K, that is we first subtract the average intensity from each pixel in each frame, then the first eigen-vector will not be the mean intensity frame (the means of the pixels are now all zero, but remember that the eigen-vectors are all orthogonal and have non-zero length). Instead the most significant eigen-vector will capture the next most salient statistic that expresses most of the variation in the data. Whatever that may be.

Another important consequence of the orthogonality between the eigen-vectors is that we can project data into the space spanned by them, simply by computing the dot-product between the data vector and each eigen-vector. Thus the coefficients a_i that express frame d as a linear combination of the eigen-vectors x_i can be found by computing the dot-products between d and the eigen-vectors.

$$d = a_{MN-1}x_{MN-1} + a_{MN-2}x_{MN-2} + \cdots + a_1 x_1 + a_0 x_0,$$

where $a_i = d^T x_i$.

Don't try this at home, though. Remember that D is of size $(M)(N) \times (M)(N)$. Television-type quality video frames are larger than $500 \times 300 =$

150, 000 pixels. There is no need to try to find the eigen-vectors of the matrix C of size $150,000 \times 150,000$. If each entry of C is a floating point number, i.e., it requires 4 bytes, then the entire matrix will need $4 \times 150,000 \times 150,000 \approx 90GB$ of space. The eigen-analysis calculation will want all of that data in memory at the same time. Admittedly, a few years from now, this may not pose a technological challenge any more, but remember that then we may repeat this argument by replacing NTSC resolution with HDTV resolution (2000×1000 pixels). A similar argument follows from the fact that we may generally not want to base classification merely on pixel intensity, as we assumed thus far. Adding further features to describe the data will lead to further increases in space requirements.

Fortunately, we can save us a lot of space and time if we rephrase our problem slightly. Let's define $C = D^T D$ instead of $C = DD^T$, then

$$Cx_i = \lambda_i x_i$$
$$D^T Dx_i = \lambda_i x_i$$

Pre-multiplying, the above by D, we get

$$DD^T(Dx_i) = \lambda_i(Dx_i)$$

where we used the facts that matrix multiplication is associative and that the λ_i are scalars. Since $C = DD^T$, Dx_i are the eigen-vectors of C.

This can make a huge difference. Remember, that D is of size $(M)(N) \times k$. As we stated above this makes DD^T a matrix of size $(M)(N) \times (M)(N)$. Likewise, $D^T D$ is a matrix of size $k \times k$. If $k << (M)(N)$ then we may save ourselves lots of compute cycles if we compute the eigen-vectors of $C = D^T D$.

This global or class eigen-analysis of the covariance matrices is useful to *describe* the features but it is not designed to *classify* feature vectors into their corresponding classes and hence often performs poorly as a basis for classification. Recent successful applications of eigen-analyses for classification, e.g., [91], normalize images of training and test objects prior to the classification to ensure that differences in the description are indeed due to differences in the identity of the objects. For instance, for the training set of a face recognition application it is necessary to crop face images manually to remove unrelated foreground and background regions, such as shades, hair, etc. Furthermore, the cropped face images need to be warped to a standard "face size", and for robustness the training set should also contain a few views of each face from a few different viewing angles, and under different lighting. [9, 10] has shown how a small number of views and lighting condition combinations can be combined to synthesize views from interpolated viewing angles and lighting conditions. While this approach works in the absence of occlusion (and self-occlusion), it does not offer a solution for natural objects. It is unclear how to standardize/normalize the size or shape of natural objects, or how to remove irrelevant foreground or background objects in their vicinity.

Finally, to achieve good performance eigen-analysis based classifiers usually require that test objects are being presented under very similar conditions as the training set. For many objects this proves to be a difficult task unless they have not already been segmented or recognized. As we said before, shifting the burden from the object recognition problem to a segmentation problem, does not get us any closer to a solution if segmentation requires detailed object information.

In classification, therefore, we are mainly interested in finding the features that maximize the *differences* between the individuals of the different classes. For this purpose the Most Discriminating Features can be obtained by maximizing the ratio $W^{-1}B$ of the **between B** and **within** class **W** sums of squares and products matrices (details in [12, 49, 78, 98]). Unfortunately, this method degenerates to Fisher's Linear Discriminating Function in the two-class case. Therefore, the difficulties of Fisher's LDF with the inverse of near-singular matrices are inherited by this eigen-analysis of the class differences.

6.   Minimally Correlated Features

Considering the problems with near-singular correlation/covariance matrices one could asked the question: *How powerful are features that allow the stable calculation of the inverse of the correlation matrix of all features (hence enabling classification using conventional linear or quadratic classifiers)?*

The answer is: *Not very;* Feature sets for which linear classifiers can be used perform roughly as well as equally large feature sets that were randomly selected from the entire set of features. The reciprocal of the condition number of the correlation matrices can be used to estimate its relative distance to the set of singular matrices. The condition number is computed using the 1-norm LINPACK condition number estimator.

The left graph of Figure 3.3 shows the performance of feature sets with the least singular covariance matrix. The error on the y-axis is measured on the training set. The average performance of random feature sets of sizes 1, 3, 6, 11, 21, and 51 are shown on the same graph as dotted lines.

The numbers in the right graph are the estimates of the covariance matrices' condition. Estimates near 1.0 indicate sets of (linearly) uncorrelated variables, while larger estimates approach the set of singular covariance matrices. The rapid increase in the singularity measure indicates that even combinations of the 20+ least correlated features cause the covariance matrix to become too singular to allow the stable computation of its inverse.

From the left graph, we can see that features for which linear classifiers can be used perform roughly as well as randomly selected feature sets of the same size. Dotted lines show the performance of random feature sets of the indicated

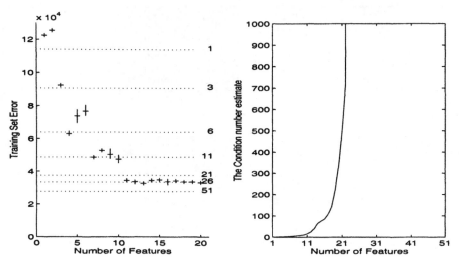

Figure 3.3. The performance of feature sets most suitable for linear analysis (left) and the corresponding estimates of their covariances' singularity measure.

size. The solid horizontal lines show the average performance of the feature sets corresponding to the least singular covariance matrices; the solid vertical lines show the performance variation due to initialization. Fortunately, we can circumvent the problems with singular covariance and correlation matrices by using methods that don't require their use (such as a back-propagation neural network). Figure 3.4 shows the performance differences between linear, quadratic and non-linear methods for a sample image.

The subset used to obtain the results of the linear and quadratic classifier in Figure 3.4 is the 20-feature set of the correlation analysis described in this section. While other sets might theoretically have better discriminatory powers, the inverses of their covariance matrices cannot be computed stably and their performance is thus unreliable (and likely to be worse).

7. Each Method in Isolation

As indicated in the Section 1 in Chapter 2, the 55 features are obtained using 7 different methods. 9 are based on color, 28 on the gray-level co-occurrence matrix, 4 on the fractal dimension, 4 on Gabor filters, 1 on the Fourier transform, 3 on steerable filters, and 6 on entropy. Since any single *feature* on its own has very little discriminatory power we instead compare each of the 7 *methods* in turn. To measure the performance of the various feature sets we average the classification error over 10 training sessions of a back-propagation neural network (BPNN). Each BPNN was trained for 100 iterations over the entire training set. The training set consists of 75 images

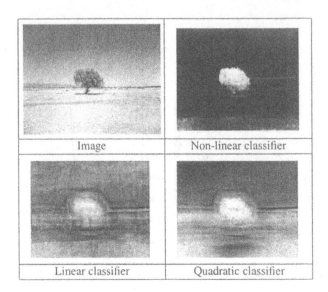

Figure 3.4. A sample image and the segmentation results obtained using different classifiers.

that were manually segmented into the classes: trees/shrub, sky/clouds, grass, human-made, animals, exhaust, and earth/rock.

Figure 3.5 shows the most relevant and surprising result of this analysis: Each method in isolation is less powerful than random feature sets of the same size. The 28 gray-level co-occurrence matrix based features always performed worse than 28 features selected at random from the entire set of 55 features. Similarly the 9 color-based measures performed worse than sets of 9 randomly selected features, and the 6 entropy-based measures performed worse than sets of 6 randomly selected features in every one of over 100 experiments. The remaining 4 sets of features, the ones based on the Fourier transform, fractal dimension, Gabor filters, and steerable filters, performed statistically worse than random sets, but due to their size there were a few experiments where these small feature sets performed better than sets of equal size of randomly selected features.

When considering the numbers of features derived from each method we can see that the performance of the different methods mostly depends on the *number* of features derived from them. This is the main reason for the apparent strength of the 28 GLCM based features. Comparing their performance with that of random feature sets of similar size we can see that the GLCM features are inferior. The numbers in Figure 3.5 indicate the number of features derived from the corresponding method, indicated on the x-axis. Single-method feature sets generally perform worse than random collections of features from the whole set. Since the goal of our analysis is to find good subsets and to optimize

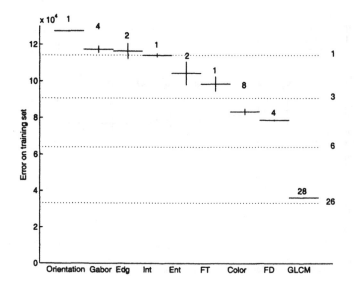

Figure 3.5.   Performance of each feature extraction method in isolation.

classification we can see that focusing on features from just one feature extraction method is not very useful. It illustrates a general problem of classification, though. Since we are interested in the best classification given *all* features we cannot judge the importance of features and methods by measuring their power in isolation.

8.   Leaving One Out

It would therefore make more sense to consider a large set and to measure how much the performance of the entire set of features deteriorates as features are left out. Since the presence or absence of a single feature is virtually unnoticeable we measure the change of the performance if all features derived from one method are omitted. The experiments were performed in the same way as described in the previous section. Although from Figure 3.6 it seems that the omission of the GLCM based features causes the largest deterioration in the error rate, this effect is almost entirely due to the large number of features based on that method. The performance of random sets shows that the GLCM features are neither particularly useful nor particularly redundant.

Another observation that can be made from this graph is that the elimination of some types of measures improves the performance of the feature set. While this indicates *that* the complete set is redundant it does not show *which* features to omit to maximize the performance. This result also points at a slight bias toward smaller feature sets. Since each training session iterated k times over the entire training image set it is not too surprising that neural networks trained

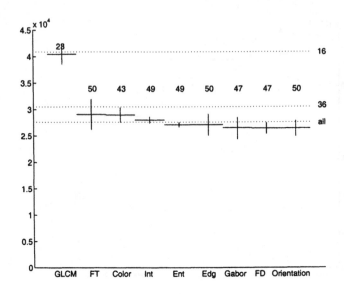

Figure 3.6. Performance of the features from the Leave-One-Out method.

on smaller feature sets occasionally performed slightly better than those trained on larger sets. Given a fixed number of training samples, smaller feature sets correspond to lower dimensional problem spaces, and thus allow faster convergence of the neural network. This illustrates a general property of classifiers. Convergence speed can often be traded for classification performance. Classification based on a single feature will quickly converge to a relatively poor result. Conversely, classification based on a large set of features will continue to converge toward better and better solutions for a long time.

9.   Feature De-correlation

As we have pointed out above, some classifiers run into numerical problems when their training data is highly linearly redundant. In those cases decorrelating the measurements can simplify a classification task. In some cases it may even enable the construction of a classifier where the   raw and highly linearly correlated data causes severe numerical instability. De-correlating features is related to feature relevance analysis, in that it may reveal redundancy in the feature set. Of course, if we suspect that de-correlating our feature set is necessary because it is too highly linearly dependent, we should not use a covariance-based method to do so (remember that those methods tend to depend on the computation of the inverse of the covariance matrix, and that this computation becomes very unstable as the correlation between the variables contributing to the covariance increases). On the other hand de-correlating measurements can cause the creation of meaningless and un-intuitive dimensions.

Eigen-analyses, such as, eigen-faces, eigen-motions, etc., construct orthogonal, i.e., perfectly de-coupled and un-correlated axes to represent the training set. Unfortunately, except for the axes corresponding to the few largest eigenvalues, the eigen-vectors (axes) tend to be meaningless for human analysis. Hundreds of relevant dimensions (i.e., dimensions associated with significant eigen-values) may be necessary to describe important facial features such as the eyes, ears, nose, and mouth. But there is no "left eye" vector that can be used to describe the variations in the left eye in face images. Even if the eigen-face analysis is done on tiny images of size 10×10 pixels, then there are up to 100 vectors that contribute to the appearance of the left eye in each face image. It is often impossible to attach any kind of semantics to the eigen-vectors corresponding to the smallest eigen-values.

Also note that eigen-analyses may well manage to reduce the feature space for a classifier by focusing on the k most significant eigen-vectors, but this may not let us prune our feature set at all. Any eigen-vector, whether it is the most or least significant eigen-vector, may depend significantly on each of the features.

The following example shows how data can sometimes be de-correlated by re-arranging it along alternative, yet meaningful axes. Figure 3.7 shows an image, and its intensities in the red, green, and blue bands.

The high correlation between the red (R), green (G), and blue (B) bands is clearly visible from the similar appearance of the intensities of the RGB color bands. The covariance matrix of the RGB representation of this image is shown here.

$$Cov(RGB) = \begin{bmatrix} 1452.5 & 1368.5 & 1328.2 \\ 1368.5 & 1713.8 & 1943.4 \\ 1328.2 & 1943.4 & 2484.0 \end{bmatrix}$$

The condition number associated with this image's RGB representation is: **161.0741**

Figure 3.8 shows an image (a), and its hue (b), saturation (c), and intensity (d).

The hue (H), saturation (S), and intensity (I) representation appears less correlated. The covariance matrix of the HSI representation of this image is shown here.

$$Cov(HSI) = \begin{bmatrix} 8554.0 & -2893.3 & 2776.5 \\ -2893.3 & 3283.7 & -2698.4 \\ 2776.5 & -2698.4 & 3538.8 \end{bmatrix}$$

The condition number associated with this image's HSI representation is: **16.3826**

The condition number measures the ratio of a matrix's largest eigen-value to its smallest eigen-value. A small eigen-value indicates that the original matrix

Figure 3.7. The red, green, and blue bands of an image are highly correlated

can be accurately reconstructed without the contribution of the associated eigen-vector. The eigen-decomposition has found a rotation of the original matrix that reveals clearly that one of its rows is little more than a linear combination of the other rows. The eigen-decomposition offers a robust method to reveal this linear redundancy in the original matrix. The condition numbers of the two matrices indicate numerically that the covariance of the HSI matrix is less singular than the covariance of the RGB matrix. This reflects the common observation that the correlation between the components of the HSI representation is smaller than the correlation between the components of the RGB representation. Note that the greater correlation between the R, G, and B color bands in Figure 3.7 is visually noticeable when compared to the hue, saturation, and intensity bands of Figure 3.8.

Figure 3.8. The hue, saturation, and intensity representation of an image are visibly less correlated than its equivalent red, green, and blue representation shown in Figure 3.7

10. Good Features and Classifiers

Efficient subsets of features generally contain representatives of all seven types of feature extraction methods. We found that combinations of features of different methods clearly outperform features based on only one feature extraction method (e.g., using only Gabor filters). We also found that back-propagation neural network classifiers perform better than convolutional neural network, linear, and quadratic classifiers. We will start the next chapter by showing that a greedy algorithm can be used to select features from the entire feature set yields small subsets that perform better than those constructed using linear or quadratic feature redundancy tests.

Figure 4.6 The blue estimate, and their color representation as an image as a widely spaced.

10. Good Features and Classifiers

The seven subsets of features generally contain representatives of all seven types of feature extraction methods. We found that combinations of features of different methods is clearly superior in feature based on only one feature extraction method is using only Gabor filters? We also found that back-propagation neural network classifiers perform better than conventional neural network, linear, and quadratic classifiers. We will show in the next chapter by showing that a greedy algorithm can be used to select features from the entire feature set yielding small subsets that perform better than those constructed using linear or quadratic feature redundancy tests.

Chapter 4

RESULTS

In this chapter we show the results of our comparison of the linear, the quadratic, the convolutional neural networ, and the back-propagation neural network and demonstrate a method to extract powerful subsets of the features used to describe still images and video frames. Good feature sets can be found that preserve much of the robustness of the entire feature set using only about a quarter of all features. Sections 3 and 4 in this chapter show the performance of the various intermediate components of the framework as well as the final classification and event detection results.

1. Comparison of Classifiers

Figure 4.1 shows a training image and the classification results of four classifiers. The classifiers were all trained on detecting deciduous trees, for which they were asked to produce white labels. Image regions that are unlikely to represent deciduous tree areas were to be labeled black. The gray-values of the result images represent   the confidence of the classifiers: a linear classifier (top center) and a quadratic classifier (top right), a convolutional neural network classifier (bottom center), and a back-propagation neural network classifier (bottom right). The performance of the two neural network classifiers appears to be roughly equivalent, while the linear and quadratic classifier can't even achieve good results on this *training* image. The neural network classifiers were trained on the entire set of features discussed in Section 1 of Chapter 2. The linear and the quadratic classifiers were trained on the largest subset of features we could find without making the subset too singular (see Section 6 in Chapter 3). Since the texture measures are based on image regions rather than pixels we have insufficient information at the borders of the image and thus the classifiers do not produce classification labels at the edges of images. Therefore, the output images are somewhat smaller in size.

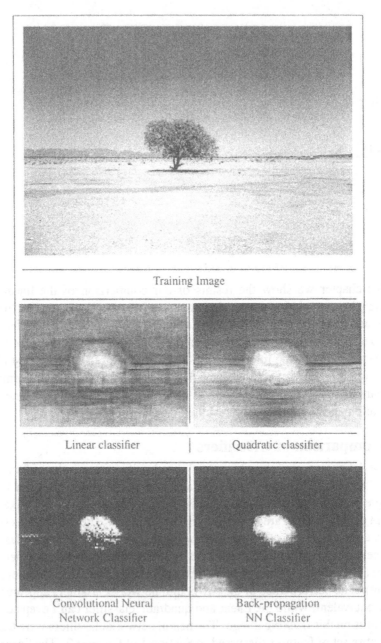

Figure 4.1.   A comparison of the performance of different classifiers on a training image.

Figure 4.2 shows that the good performance of the CNN is due to over-training, i.e., memorization of the training data. The performance of the convolutional neural network on this test image shows that it does not generalize well on unseen images (classification results in center). The back-propagation neural network classifies the image regions much better (classification results

Training Image

| Convolutional Neural Network Classifier | Back-propagation NN Classifier |

Figure 4.2. A comparison of the performance of different classifiers on a training image.

on right). Furthermore, both training the network and testing/labeling images required more time than for the approach with separate preprocessing and labeling stages for the back-propagation neural network. One of the motivating factors in the design of CNNs is to "wire" some of the domain knowledge into the network architecture to improve performance. In [72] a CNN is used for handwritten digit recognition. To reduce the number of weights in the network all units at the same level use a common set of weights (a convolution kernel) to connect them to the units of the next lower level of the network. Our comparison shows that this is not enough to obtain good classification results. Training a classifier on pre-processed input therefore seems to be the better approach. Therefore, we chose to use the back-propagation neural network classifier for our object classification task.

2.   Selecting Expressive Subsets of Features

Selecting subsets of features from a given set is generally a difficult problem, as the number of subsets can be prohibitively large even for moderate numbers of features. To establish a lower bound of performance for subsets of varying size we first averaged the performance of the classifier on random feature samples as described above in Chapter 3 Section 7. We assume that the discriminatory power of a set of features can be estimated by averaging over the differences due to initialization. In the following figures, horizontal lines represent averages of solutions of a given feature set and vertical lines represent the ranges of these solutions.

2.1.   Random Sets of Features

Figure 4.3 shows the errors on the training set as the number of randomly selected features is increased from 1 to all features. Horizontal lines show the average error for the feature sets of the various sizes, vertical lines indicate the range of errors for each feature set.

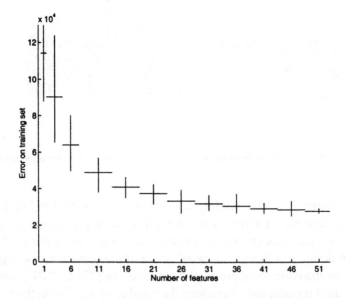

Figure 4.3.   The performance of random feature sets of various sizes.

These results should form a lower bound on the performance of any feature selection method. Difficulties with the correct label for each pixel, problems with pixels on the border between tree and non-tree regions and problems due to compression prevent the complete elimination of the error. This, partly, accounts for the fact that the exponential error curve does not approach zero as the number of features is increased. This is not to say that, using additional/alternative methods, we cannot reduce the error further even with smaller numbers of features.

2.2. Good Subsets of Features

Transitivity does not hold for subsets of features: given an optimal subset with N features we cannot expect that the addition of the "best" feature not yet in the set will produce an optimal set of size $N + 1$. Likewise, eliminating the "least useful" feature from an optimal set of size $N + 1$ need not produce the best subset of size N. Therefore, for N features an optimal solution needs to consider 2^N subsets.

2.3. A Greedy Algorithm

If we use a greedy algorithm that starts from the empty set and includes the single most useful feature at each step (or its complement, a greedy algorithm that starts with the entire set and discards the single most useless feature at each step), the resulting subsets of features are generally no better than randomly selected sets of the same size. This somewhat surprising result is due to the fact that the greedy algorithm will make locally optimal choices. After a few iterations (during which additional features are being added), they turn out to be worse than if we had picked the same number of features at random.

2.4. Beyond the Greedy Algorithm

Starting a greedy algorithm from good random feature sets we can typically observe a performance increase during the first few steps, which goes beyond that expected from equally sized random feature sets. After a few steps of the algorithm, though, the performance increases diminish and the performance again approaches that of randomly selected sets.

Since the performance of a set of features varies strongly with the initialization of the search we average the performance of a few differently initialized searches. We used the best of 10 random sets of 6 features as a starting point of an incremental greedy algorithm. To broaden the horizon of the search we also checked to see if deleting a feature or replacing one feature with another yields better set performance.

Using this strategy we determined 13 features that achieve an error rate on the training set of 28373, which is only about 3 percent worse than the collection of all 55 features combined (27500). The execution time of the feature extraction stage, on the other hand, is reduced to less than one quarter that needed to extract all 55 features. From Figure 4.3 it can be seen that the performance of these 13 features is roughly equivalent to the performance of the entire feature set. Figure 4.4 illustrates the steps of the method for a particular set of features.

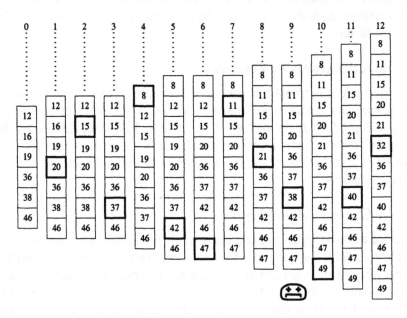

Figure 4.4.   13 steps of the greedy algorithm, starting from a set of six randomly selected features.

The initial 6-feature set is shown in column 0 of feature IDs. In the first step it was found that feature 20, which was not in the 6-feature set reduced the error the most among all features not in the 6-feature set. The resulting 7-feature set is shown in column 1. In the next two steps features 16 and 38 were replaced by features 15 and 37, respectively. In step 4 feature 8 was found to be the best addition to the existing set. In steps 6 and 7 features 19 and 12 were replaced by features 47 and 11, respectively. Steps 8 and 9 did not produce clear enough results and so two alternative feature sets were kept and further expanded. At the next step (10), however, it turned out that reintroducing feature 38, which was replaced with feature 37 in step 3, was not as beneficial as adding feature 21 to the set. The "sad face" marks the elimination of an alternative feature set. Finally, in steps 11 and 12 features 40 and 32 were added to the set of features.

Figure 4.5 compares the performance of random feature sets from Figure 4.3 and the performance of the feature sets shown in Figure 4.4 (the circles in the graph). As mentioned earlier, it can be seen that the performance-increases taper off as the number of steps of the greedy algorithm increases. The final

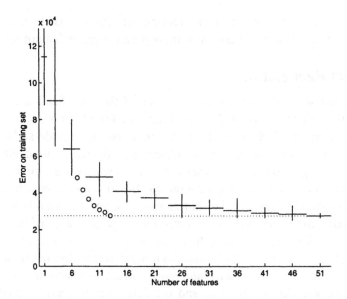

Figure 4.5. The circles show the average performance of seven feature sets obtained starting from the best 6 feature set (the lower tip of the vertical line corresponding to the performance range of random feature sets of size 6).

set of 13 features has the following components: **The first 5 features of the best reduced feature set are based on the GLCM, the 6th is a Gabor filter measure, the 7th a multi-fractal measure, the 8th, 11th, 12th and 13th are based on color measurements, the 9th is based on entropy, and the 10th on the Fourier Transform.** With the exception of the steerable filter based features all types of measures are represented. This confirms the need for the combination of features of different type for good classification.

The main reason for the inclusion of the steerable feature measures was to facilitate the detection of branches, forks, and junctions in images (we did not use these features in this work). But we also used the orientation, and edge type information to see if they could contribute directly to the classification process, rather than just indirectly through the detection of branches and forks. The absence of these features from the final 13-feature set indicates that without further processing they are not very useful.

The performance of the final set of features is close to that of the entire feature set, when the neural networks are trained for the same number of iterations and performance is measured on the training set. As we said before, changing the size of the feature set biases the classification error on the training set toward smaller feature sets. Larger feature sets often enable the classifier to continue to learn from the training set long after the classifiers based on smaller feature sets have converged (to relatively worse solutions). Classifiers trained on smaller feature sets often achieved the good classification results due to over-training, i.e., memorization of the training set. As a consequence those classifiers generalize less well than those trained on the entire feature set.

3.   Object Recognition

In this section we will show the performance of the classification module. To evaluate the effectiveness of the algorithm, we tested it on a number of commercially available VHS tapes from different content providers. We will show the results of this module for deciduous tree detection application as well as hunt, landing, and rocket launch events. Then we show examples of the extracted texture and color features, the motion estimation and detection results, the region classification results, the shot summaries, and the final hunt event detection results. In this section we demonstrate the performance of the neural network classifier for the detection of deciduous trees in over 50 still images, the detection of sky and clouds in 20 frames from unconstrained video sequences the detection of sky, clouds, exhaust, and human made structures in 20 frames from unconstrained video sequences and the detection of grass, trees/shrub, sky/cloud, rock, and animals in 20 frames of wildlife documentaries. A total of 45 minutes of footage $\triangleq$ 81000 frames $\triangleq$ 270 shots $\triangleq$ 36GB (uncompressed) have been processed.

3.1.   Feature Space Representation of the Video Frames

Figure 4.6 shows the feature space representation of a video frame. The features shown are the results of the gray-level co-occurrence matrix based measures (first 56 feature images), the fractal dimension based measures (next 4 feature images), the color based measures (next 4 feature images), and the Gabor based measures (last 12 feature images).

The feature maps for the entropy, steerable filter, and Fourier transform features are not shown here. The described feature set enables the robust detection of a range of natural and human made object classes. Some of the test images in Figure 4.7, Figure 4.8, Figure 4.9, and Figure 4.10 show deciduous trees in fall with the leaves' colors ranging from green, through yellow, orange and red to a magenta-ish red. Color is often a useful cue, but the classifier has also learned that leaves are not always green and that not everything green depicts leaves. Similar observations are true for the other features; often a range of their activity corresponds strongly, but not exclusively, with deciduous trees. The task of the classifier is to fuse the measures and improve on their individual performance.

Figures 4.7, 4.8, 4.9, and Figure 4.10, each show 3 rows of 4 images with the output of the tree detector neural network classifier below them. Since there is no image data beyond the edges of the images most of the texture measures

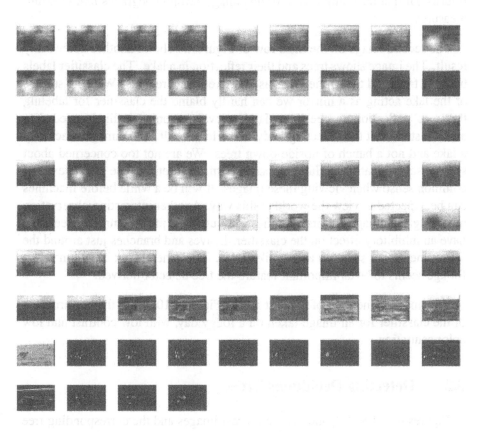

Figure 4.6. The feature space representation of the first frame in Figure 4.15.

cannot be computed around the edges. Therefore, the classifier output images are smaller than the corresponding images. The tree detector classifier output is shown below the image in gray-values ranging from white – representing pixels that are very tree like – to black – representing pixels that are very different from trees.

The third image on the third row of Figure 4.7 shows the approach's robustness with respect to scale and color. This fall image shows trees at distances ranging between 5 meters and over 500 meters whose colors range from magenta to green. The classifier output image below it shows that almost all tree regions were correctly labeled.

The third image on the first row of Figure 4.8, though, shows false positive matches on the beach where some of the sand is found to be somewhat tree/leaf like. The last image on the same row shows an example of false negatives. The classifier does not label the large tree region along the river as very tree-like.

It turns out that the compression of this image severely degrades many texture measures.

The second image on the third row of Figure 4.9 shows another noteworthy result. The image shows trees and their reflection in a lake. The classifier labels both the trees and their reflections as tree-like image regions. With the surface of the lake acting as a mirror we can hardly blame the classifier for labeling the trees' reflection as tree like. Higher-level reasoning is likely needed to assert instead that the reflecting body in the lower half of the image is actually a lake and not a bunch of upside-down trees. We are not too concerned about such misclassifications. Human observers make extensive use of context and common sense when viewing these images. It will be a while before machines can be expected to view the world in this way. Another interesting observation regarding this image and its classification output is, pthat symmetry seems to have an inhibitory effect on the classifier. Leaves and branches just around the water line do not show up as tree like areas. The general lack of symmetry in foliage seems to be an important feature the tree detector derived.

The image at the right of the first row of Figure 4.10 shows the performance of the classifier for an image taken on a foggy day, with low contrast and low color saturation.

3.2.   Detecting Deciduous Trees

Figures 4.7, 4.8, 4.9, and 4.10 show test images and the corresponding tree detection results. Test images are new, unseen images, i.e., they were not used to train the classifier. The detection results show gray-level images where white pixels correspond to tree regions and black pixels correspond to non-tree regions. The gray-tones between black and white indicate areas of varying tree-likeness.

Note that not everything green in images has been labeled as tree regions, see for instance the third image on row 5 of Figure 4.9. Although, not visible in the gray-level print of this book, the tree has rusty brown leaves that were detected to be tree-like, while the green grass in the foreground was labeled as un-tree-like.

Among the many test images, including the first one we mentioned explicitly above (the fourth image on row 1 of Figure 4.10) there are a few where accumulations of fallen leaves in the grass were labeled as tree regions. Again, we would like to argue that human observers of the same scenes would make use of much higher-level processes to correct our bottom-up hypotheses. After all, the foliage itself is indistinguishable from the foliage still attached to the trees.

Figure 4.7. Test images and the corresponding tree classification results.

Figure 4.8. Test images and the corresponding tree classification results.

Figure 4.9. Test images and the corresponding tree classification results.

Figure 4.10.   Test images and the corresponding tree classification results.

3.3. Detecting Grass, Trees, Sky, Rock, and Animals in Wildlife Documentaries

The segmentation results in Figure 4.11 show the performance of the neural network classifier on frames from wildlife documentaries. As with all other classification results shown in this book the results were obtained for images from sequences that were not used to train the classifier. Rows 1, 3, 5, and 7 of Figure 4.11 show a number of frames from hunts and their classification results (rows 2, 4, 6, and 8). The classification results are encoded as follows: blue represents sky or clouds, light green represents grass, dark green represents trees, black represents rock, and white and yellow represent animal regions.

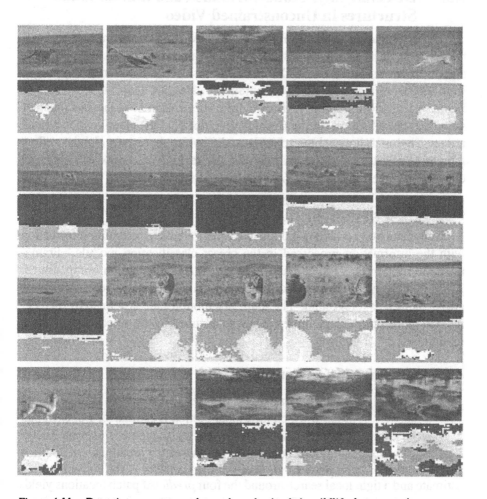

Figure 4.11. Detecting grass, trees, sky, rock, and animals in wildlife documentaries.

3.4.   Detecting Sky in Unconstrained Video

The segmentation results in Figure 4.11 the performance of the neural network classifier on frames from unconstrained video. As with all other classification results shown in this book the results were obtained for images from sequences that were not used to train the classifier. Rows 1, 3, 5, and 7 of Figure 4.11 show a number of frames from landing sequences together with their classification results (rows 2, 4, 6, and 8). The classification results are encoded as follows: blue represents sky or clouds while black represents non-sky regions.

3.5.   Detecting Sky, Clouds, Exhaust, and human-made Structures in Unconstrained Video

The segmentation results in Figure 4.13 show the performance of the neural network classifier on frames from unconstrained video. As with all other classification results shown in this book the results were obtained for images from sequences that were not used to train the classifier. Rows 1, 3, 5, and 7 of Figure 4.13 show a number of frames from rocket launch sequences together with their classification results (rows 2, 4, 6, and 8). The classification results are encoded as follows: blue represents sky, white represents clouds, yellow represents exhaust, black represents non-sky regions, and pink is a catch-all class for all pixels that are not similar enough to any class.

4.   Event Detection

Event detection combines motion and foreground segmentation information with the information processed in the object classification module. Frame-to-frame motion is estimated and adjacent frames are motion compensated. Areas that are not well-matched by the motion compensated reference frame are segmented as foreground. The shape of a motion-blob is defined by this segmentation; its object class properties are determined by the object classification module.

4.1.   Global Motion Estimation

Figure 4.14(a) shows the size and locations of the four regions at which the global motion is estimated. For each pair of frames motion estimates are computed using a 5 level pyramid scheme at the shown patch locations. In addition the previous motion estimate is taken as the current motion estimate and a tight local search around the four *predicted* patch locations yields another four patch matches. The best match of any of these 8 patch comparisons becomes the motion estimate for the current frame pair, as described in Section 3 of Chapter 2. Figure 4.14(b) shows the motion estimates during a hunt. Global

Figure 4.12. Detecting sky regions in unconstrained video sequences.

motion estimates such as the ones in Figure 4.14 (b) are used to detect moving objects as shown in Figure 4.15. The locations of these moving object blobs are then verified using a neural network image region classifier that combines color, texture, and this motion information, as shown in Section 2 of Chapter 2.

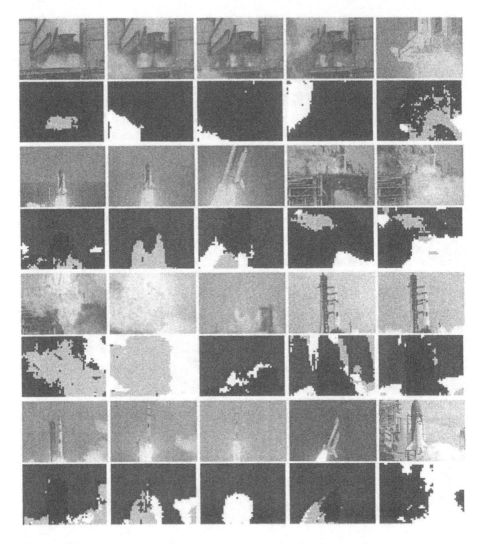

Figure 4.13.   Detecting sky, clouds, exhaust and an all-other-objects regions in unconstrained video sequences.

4.2.   Motion-Blob Detection

Figure 4.15 shows an example of the motion-blob detection results. The figure shows two consecutive frames, 4.15(a) and (b), and the difference image (c) that results from subtracting (a) from (b). Note that in (c) areas of zero difference between the two frames are represented by the medium shade of gray. Areas of positive differences are brighter, while areas of negative

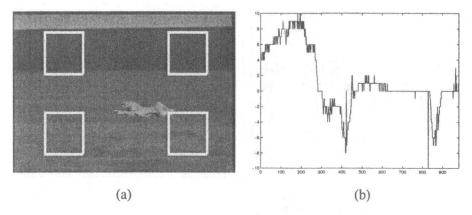

Figure 4.14. The locations used to estimate the global motion (a), and the motion estimates during a hunt (b). The plot shows the horizontal motion (y-axis) in each frame (x-axis).

differences are darker. The absolute differences between the two frames is used in the motion estimation method described in Section 3 of Chapter 2. Regions found to belong to the dominant motion-blob(s) are segmented and are labeled as foreground.

Using this approach we compensate for the detected motion before computing the difference between the two frames. The steps of this approach are shown in Figure 4.15. Reliable estimation and compensation of global motion can be seen to simplify the task of motion-blob detection. When the accuracy of the global motion estimation results are poor, the performance of the motion-blob detection relies largely on the robustness of the motion blob detection and tracking algorithm described in Section 4 of Chapter 2.

4.3. Shot Summarization

The intermediate level process consists of two stages. In the first stage the global motion estimates are analyzed and directional changes are detected in the x and y directions. When the *signs* of the 50 frame global motion averages before and after the current frame differ and their *magnitudes* are greater than 1 pixel per frame we insert an artificial shot boundary. Artificial shot boundaries are also inserted every k frames so as to enforce some amount of regularity on the shot summary. The purpose of this rule is mainly to ensure that shot summaries do not summarize vastly different shot durations. In the second stage each shot is then summarized as in the example shown below. We do not show the results of the shot boundary detection module since the results are comparable to those of many other popular boundary detection methods. The module achieves about 80% correct shot boundary detection and

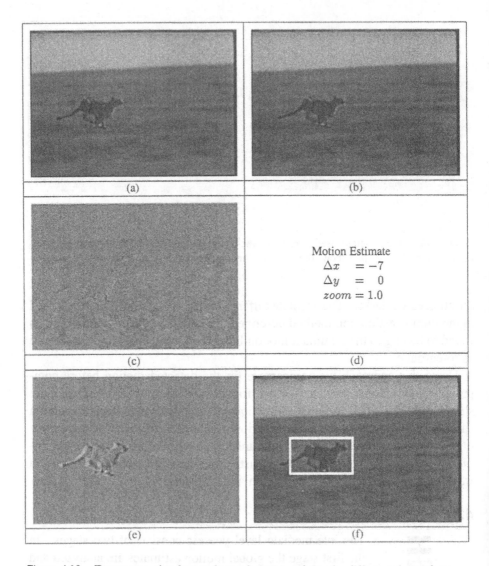

Figure 4.15. Two consecutive frames from a hunt (a) and (b), the difference image between them (c), the estimated motion between the two frames (d), the motion compensated difference image (e), and the box around the area of largest residual error in the motion compensated difference image.

suffers the same problems as other shot boundary detection schemes when shots boundaries are dissolves, long overlays, cross fades, or other special effects and transitions.

4.4. Event Inference and Final Detection Results

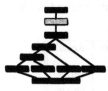

To demonstrate the usefulness of our approach to visual event detection the presented method is used to detect hunts in wildlife documentaries as well as landings and rocket launches in unconstrained video. The next three sections show and discuss the results of these three applications.

4.5. Hunt Detection

The event inference module infers the occurrence of a hunt based on the intermediate descriptors as described in Section 4.3 of Chapter 4. In doing so, it employs four predicates, Beginning of hunt, Number of hunt shot candidates, End of hunt, and Valid hunt, which are currently embedded in the shot summary. If the intermediate descriptors Tracking, Fast and Animal are all true for a given shot, the inference module sets Beginning of hunt to be true, which means the shot could potentially be the beginning of a hunt event. The inference module tracks the intermediate descriptors Tracking, Fast and Animal for consecutive shots and increments the value of the Number of hunt shot candidates if all those three descriptors hold true for consecutive shots. In our current implementation, when the Number of hunt shot candidates is equal or greater than 3, Valid hunt is set to be true. Finally the inference module sets End of hunt to be true if one of the intermediate descriptors Tracking, Fast and Animal becomes false, which implies either the animal is no longer visible or trackable, or the global motion is slow enough indicating a sudden stop after fast chasing.

In the results, hunt events are specified in terms of their starting and ending frame numbers. There are 7 hunt events in the 10 minutes (18000 frames) of wildlife video footage, which were processed. Table 4.1 shows the actual frame numbers of the 7 hunts and all the frames of the detected hunts when we applied the presented algorithm to the 10 minute video footage. The table also shows the retrieval performance of the presented method in terms of the two commonly used evaluation criteria (1) precision and (2) recall.

The recall and the precision measures for hunt 6 are bad because both the predator and the prey disappeared behind a hill for a few seconds during the hunt. Only when both are later seen at the other side of the hill it becomes clear that the hunt is still ongoing. The long occlusion of both animals caused the event detector to be reset, since no animals could be tracked for a number of shots. While this result indicates a limitation of our event detector it is difficult to envisage a solution for this kind of problem. It is likely that some kind of spatial model of the scene needs to be built from the video to understand that the animals are first hidden by the hill and then reappear behind it.

Table 4.1. A comparison of the actual and detected hunts in terms of the first and last hunt frame, and the associated precision and recall.

Sequence Name	Actual Hunt Frames			Detected Hunt Frames			Precision	Recall
hunt1	305	-	1375	305	-	1375	100 %	100 %
hunt2	2472	-	2696	2472	-	2695	100 %	99.6%
hunt3	3178	-	3893	3178	-	3856	100 %	94.8%
hunt4	6363	-	7106	6363	-	7082	100 %	96.8%
hunt5	9694	-	10303	9694	-	10302	100 %	99.8%
hunt6	12763	-	14178	12463	-	13389	67.7%	44.2%
hunt7	16581	-	17293	16816	-	17298	99.0%	67.0%
Average							95.3%	86.0%

Figure 4.16 shows eight frames from the short hunt in sequence 7. Most of the 1800 frame ($\stackrel{\triangle}{=}$ 1 minute) sequence is either unrelated to any hunt or can at best be interpreted as stalking. Between 5 and 7 seconds of the sequence are sufficiently hunt like. At times the animal is barely visible between the scrub and trees. The bad visibility and the fact that the average true motion during the hunt shots is over 30 pixels made this a tough test, and we are pleased to see that the short hunt was indeed detected by the program. The detected hunt event stretches from the fourth shown frame to the last.

For the remaining events we will show key-frames indicating the transition between the stages of the event detector. Since for hunts there are no meaningful intermediate stages we cannot show such intermediate information for this application. For the remaining events, such intermediate information will be shown.

4.6.   Landing Events

Landing events can be broken into three phases.

Approach: Initially the tracked, descending object is seen with sky below it.

Touch-down: Following the Approach the tracked, descending object can be seen to have a large horizontal motion component.

Deceleration: A sudden reduction of the horizontal speed of the tracked and now grounded object represent this final phase of the landing event.

Figures 4.17 and 4.18 show these three phases for 6 landing sequences. The phases of the 6 landing events were correctly detected in all but the last landing sequence. Detection errors can usually be traced back to the failure of one of the modules in our framework. In the case of the last landing sequence

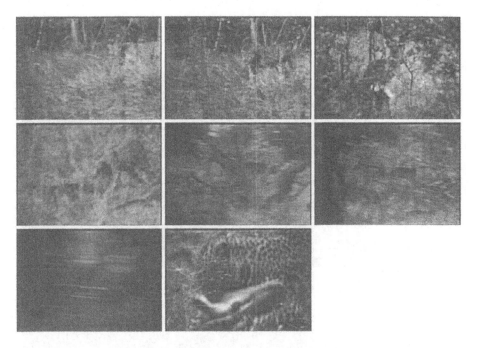

Figure 4.16.   Every 30th frame from a short hunt in a wildlife documentary sequence.

in Figure 4.18 only the approach and touch-down phases were found. The deceleration phase could not be found since the frames following the landing phase fade to black before the aircraft slows down sufficiently, as shown in Figure 4.19. Fades, sunsets, rain, surf, or rivers are often used to indicate a context switch or the passing of time. Since the focus of this book is on visual object and event detection we have not attempted to capture or interpret editing patterns such as the fade-to-black shown in this video sequence. The (soothing) effects that sunsets, surf, etc. have on humans are exploited by moviemakers to convey important information, but it is difficult to interpret such information automatically with an artificial system that has no concept or representation of excitation and calmness. In the absence of the deceleration phase the event detection module at the highest level of the prsented framework does not assert the detection of a landing event.

Figure 4.20 shows a frame from a sequence for which the event detector failed because the classifier misclassified the salt lake on which the space shuttle is landing as a sky/cloud region in the sequence. Due to this misclassification the landing detector located sky below the motion-blob and prevented the control from moving to the landing state (since for this state the moving object must touch down on a non-sky part of the image). It is unclear whether humans can visually determine that the space shuttle is rolling on a salt lake or flying above

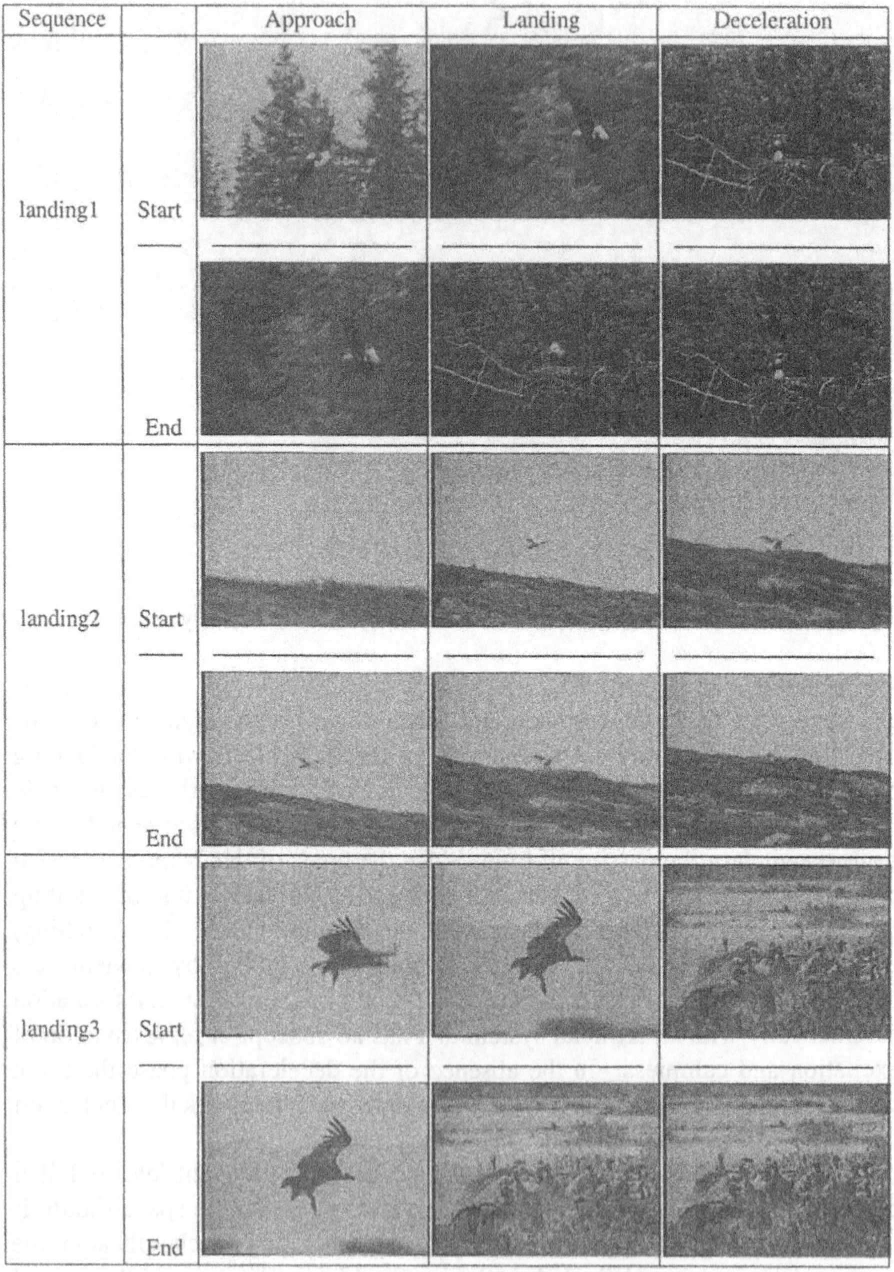

Sequence		Approach	Landing	Deceleration
landing1	Start			
	End			
landing2	Start			
	End			
landing3	Start			
	End			

Figure 4.17. The detected phases of 3 landing events.

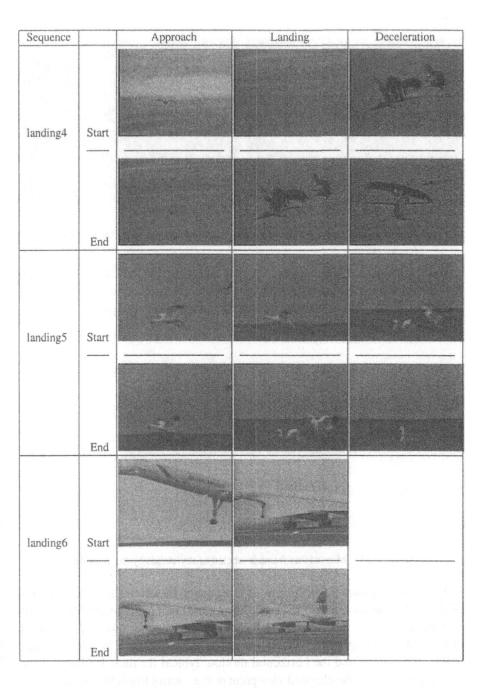

Figure 4.18. The detected phases of 3 more landing events.

Figure 4.19.   Before the aircraft slows down significantly the video fades to black.

a uniform white cloud. Removing the space shuttle from the frame (as done on the right of Figure 4.20 makes it obvious that the classification task is difficult without context, common sense, and/or background knowledge.

Figure 4.20.   Most of this frame is labeled as sky/clouds by the classifier.

Figure 4.21 shows a few frames of a sequence in which the characteristics of the landing event are very different from most landings. Initially the aircraft is seen in the sky. The camera is located in the control tower of an airport close to the landing strip. As the aircraft is approaching it is largely flying toward the camera and the horizontal motion, typical for most landing events, is negligible. From the elevated viewpoint of the control tower the aircraft drops below the horizon before its horizontal motion becomes significant. Since the aircraft was not detected as a flying object early on in the sequence the event detector control does not continue to track it near/on the ground even when the

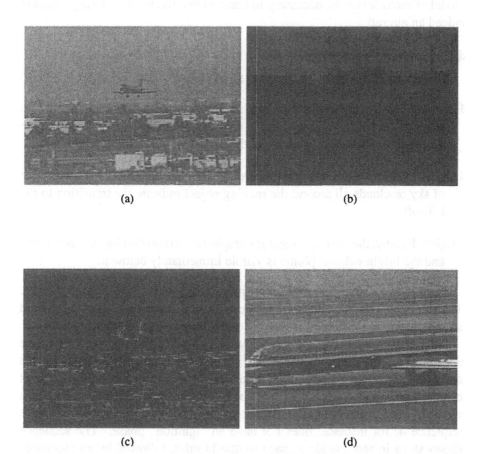

Figure 4.21. The frame in (a) shows a landing aircraft just above buildings at the end of the runway. The resolution of the classification output for the frame in (a) shows the aircraft as part of the ground (b). Note that the output of the classifier does not yield results at the edges of the frame. (c) shows the residual error after compensating for the estimated background motion in frame (a). The large area of moderate residual background error masks the error due to the independent motion of the aircraft in the frame. The frame in (d) shows that due to an aperture problem the fast motion of the aircraft cannot be recovered after it has touched down on the runway. The combination of these problems prevented the event detector from finding this landing event.

horizontal camera motion finally does pick up. At this point the only difference between a landing aircraft and a car or truck is the shape of the object, which we do not model explicitly. Therefore, this landing event remains undetected.

To avoid this problem (a) the motion detection module's performance needs to be improved to detect the flying object while it is in the air, (b) a more elaborate model of aircraft may be necessary to confirm that the tracked, flying object is indeed an aircraft.

4.7.   Rocket Launches

Rocket launches can be broken into three phases.

Ignition: Initially the engines are started, bright exhaust becomes visible, and clouds of exhaust begin to form. The only noticeable motion is due to the non-rigid deformation (growth) of these clouds.

Lift-off: After the ignition the onset of an upward motion and the presence of sky or clouds all around the moving object indicate the transition to the Lift-off.

Flight: Finally, the moving object is completely surrounded by sky and clouds and the bright exhaust plume is visible immediately below it.

Figures 4.22 and 4.23 show the three phases for 7 rocket launching sequences. The phases of 7 rocket launch events were correctly detected except in the third sequence, where the correct and an incorrect launch were detected.

All 8 launch events in the test set were correctly detected, with one false detection during launch sequence 3a. Furthermore, the rocket launch phases detected in sequences 1 and 7 closely match the phases of the depicted launch events. The fact that not all the launch phases of the remaining video sequences were detected correctly has a number of reasons and implications. Launch sequence 4, for instance, does not have an "ignition" phase. The sequence shows shots in and outside a space shuttle in orbit, followed by an ascending shuttle after its lift-off, followed by further shots outside the orbiting shuttle. Since the rocket launch model in Figure 2.17 does not state the presence of non-sky/clouds/exhaust regions below the rocket during the ignition phase, the appearance of exhaust in the first launch shot of the sequence is mistaken for the ignition phase of the launch. Sequences 2, 5, and 6 show that it is not necessary to detect the exact locations of the boundaries between the phases of rocket launch events to detect the event as a whole. In sequence 2 the beginning and end of the lift-off phases was detected incorrectly. In sequence 5 part of the ignition sequence was missed and in sequence 6 the lift-off phase ends prematurely. In launch sequence 3a a false rocket launch was detected for two reasons, 1) skin color was mistaken for exhaust (largely because the training set for exhaust did not contain negative examples showing human skin), and 2) the motion estimation failed (largely due to the multiple motions of the people in the sequence, which violates our assumption of a uniform background motion).

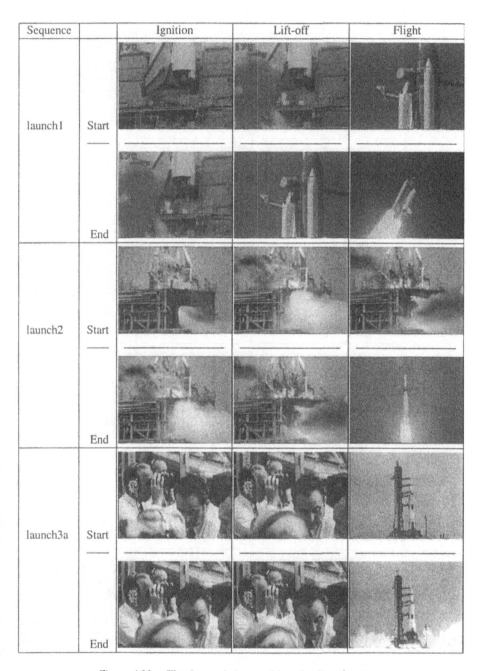

Figure 4.22. The detected phases of 3 rocket launch events.

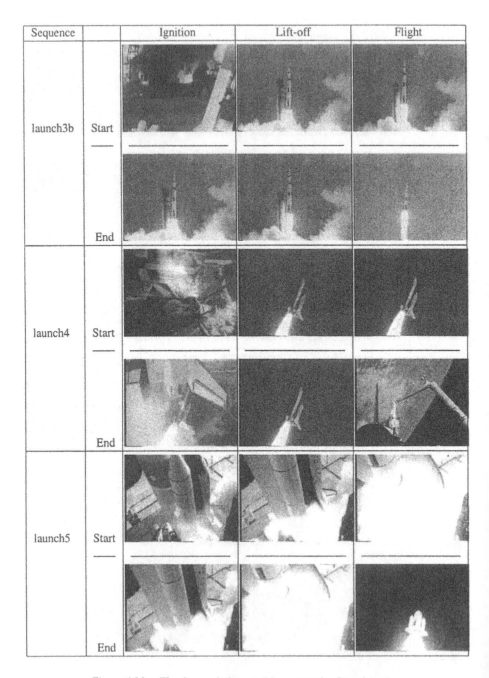

Figure 4.23. The detected phases of 3 more rocket launch events.

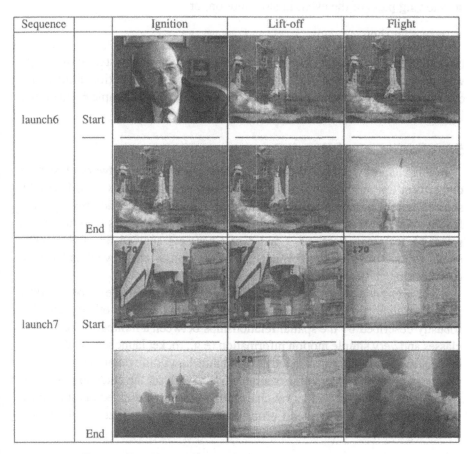

Figure 4.24. The detected phases of 2 more rocket launch events.

However, the correct launch event was still detected, albeit with a slightly shortened "ignition" phase, as shown in launch sequence 3b.

This indicates that the detection of rocket launches is rather straight forward given the classification of image regions into sky, cloud, exhaust and human-made/structured object regions. Humans easily combine the presence of exhaust below an ascending object surrounded by sky or clouds to infer a rocket launch, even from still images. A missing launch phase or the imprecise detection of its beginning or endpoints can be compensated for by the information gathered in the other phases.

Further redundancy in rocket launch events is added when a number of sources are edited into a single launch sequence. The relatively short launch process is often artificially stretched by

- showing a number of different views of the same visual event,

- showing parts of the event in slow motion, or

- repeating shots.

Not only can the well-developed human visual recognition system easily cope with these distortions but slight challenges to our pattern detection abilities are even considered good film-making practice and often make simple events more interesting to watch.

5.   Summary of Results

In this chapter we showed that back-propagation neural network classifiers outperform linear, quadratic, and convolutional neural network classifiers. We showed how a greedy algorithm seeded at a good set of randomly selected features can be used to obtain a powerful subset of 13 features from the entire set of 55 features. Sections 3 and 4 of Chapter 4 showed the performance of the various intermediate components of the framework as well as the final classification and event detection results.

The 3 presented events indicate that some events are easily described by little more than their motion characteristics (e.g., landing events), other events are robustly described by the spatial relationships between a number of key object classes and very simple motion information (e.g., rocket launches), and yet others require a mix of both, as in the case of hunts in wildlife documentaries. The presented approach, thus, offers a simple framework for the detection of events with typical spatial and temporal characteristics. The presented automatic extraction of spatial and temporal primitives can easily be utilized to describe a wide range of such events.

Chapter 5

SUMMARY AND DISCUSSION OF ALTERNATIVES

We have presented a new computational framework and a number of algorithmic components that enable automatic object recognition and event detection and applied it to the detection of deciduous trees in still images, hunts in wildlife documentaries, and landings and rocket launches in general unconstrained commercial video. Before summarizing these topics we will turn to issues concerning the choice of the classifier and the feature set.

1.   Classifiers and Features

This section summarizes the observations of the experiments with different feature sets and types of classifier, and it offers some useful insights into issues that often arise during classification. Gathering meaningful descriptions of images and video is the first step toward robust object/event detection and recognition. Invariably, we will be tempted to rid the resultant feature set of some of its redundancy. A lean but expressive feature set helps us better understand the classifier's decision. We might even want to try to hand craft a decision tree using a few of the most useful features. Decision trees are trees of decision functions. In every branch and at every level a decision tree uses a decision function to determine along which branch to descend next. Consider the following simple decision tree for deciding whether a pixel represents a part of a deciduous tree. We realize that classifying deciduous trees using a decision *tree* algorithm may be confusing, but deciduous trees serve as an excellent example that represents many real world classes, and exhibits many real world problems.

Since deciduous trees have green leaves for a large part of the year, we might first ask whether the pixel is green. If it is we might next want to ask whether it has very dark or very bright neighbors, expressing our observation that foliage often exhibits bright specular reflections next to dark regions due to shadows cast by leaves on other leaves, or due to a gap in the canopy.

On the other hand, if a pixel's color is not green, but yellow, we might still be looking at a part of a tree. It might be an image of a tree in fall. We may guess that if this were true then there should be other fall colors nearby, such as green, yellow, or red pixels belonging to neighboring leaves.

Our little 2-level decision tree consists of three questions. At the first level we ask whether the pixel in question is green or not. At the second level we either check whether neighboring pixels are very bright or very dark, or in case we answered *No* to the first question, we ask whether it has fall-colored neighbors. When we reach a "leaf" of the decision tree we can read off the label for the pixel of interest. For our 2-level decision tree we might assign the following class labels to the "leaves" of the decision tree: If we answered *No* to both questions, we may conclude that the pixel does not belong to a deciduos tree region in the image, because the pixel is neither green, nor does it seem to have fall-colored neighbors. If we answered *No* to the first question and *Yes* to the second we may assert that the pixel belongs to a deciduous tree region in the image. If we answered *Yes* to the first question and *No* to the second we may assert that the pixel is part of a smoothly varying part of the image and thus is unlikely to represent a deciduous tree. Finally, if we answered *Yes* to both questions then we may assert that the green pixel whose neighbors' brightness varies strongly belongs to the deciduous tree region.

This little 2-level decision tree shows some of the fundamental problems with decision trees in the presence of vague and imprecise measurements, which by the way is almost always the case in real world problems. The main problem stems from the vast number of assumptions we make at every step down the decision tree.

We assume that foliage is green. But often this is not true even in summer. There are many trees whose leaves range from magenta through rusty-red, through turquoise-green. Shadows cast on objects on a sunny day often shift the colors reflected by these objects toward blue. In the absence of the direct light of the sun the blue sky becomes the next most powerful source of ambient light. This may let green leaves appear to have significant and even dominant blue components.

Of course, during fall we may not find any green leaves. In spring things may be complicated due to the fact that trees blossom in just about any color imaginable. And we may see the blue sky between the budding leaves.

Another problem with decision trees is the sequential nature of the questions we are asking and the hard decisions we make in response. If we make just one bad decision, say we find the texture in the neighborhood of a leaf to be a little too regular, symmetric, or well-structured we may descend down the wrong part of the decision tree. Now we are barking up the wrong tree. Just as in our little toy decision tree the questions asked in this part of the tree may be designed to deal with a special case of the object of interest, certain lighting conditions, an unusual viewing angle, etc.

This example should mainly serve to show that hand crafting decision trees is a difficult task. Almost any classifier can yield better results than such manually designed decision trees.

The remainder of this chapter will focus on some of the problems associated with some classifiers, features, feature sets, and feature reduction methods.

1.1. Correlation, Orthogonality, and Independence

If the values two measures take as we compute them at every pixel of every image are un-correlated, are they orthogonal? Are they independent? What if we know that two variables are independent, are they orthogonal?

Two variables X, and Y are un-correlated if

$$Cov(X, Y) = E[XY] - E[X]E[Y] = 0$$

Two variables X, and Y are orthogonal if

$$E[XY] = 0$$

Two variables X, and Y are independent if the value of X is not affected by the value of Y, in any way.

Among these three definitions independence is the strongest constraint between the variables X and Y. Any two variables that are independent are also orthogonal and un-correlated. Two variables that are not independent but orthogonal are also uncorrelated. Two variables that are uncorrelated need not be orthogonal.

The left plot in Figure 3.2 on page 66 showed an example of two variables that are uncorrelated but not orthogonal. Since $Cov(X, Y) = 0$, but neither of the two variables' means are zero (as can be seen from the plot) the two cannot be orthogonal (see the definitions for orthogonality and correlation, above).

Two variables whose data points fall on the two axes of a 2-D coordinate system is orthogonal, the variables are not independent. Whenever one of the variables has a non-zero value the other variable is zero, and vice versa. This constraint between the two variables shows that there is a rule that describes the dependence on each other. By the definition of independence there are no rules between independent variables.

1.2. Linear Dependence

Covariance (or correlation) matrix-based feature relevance analyses fail even if only a small number of features are significantly *linearly* correlated. Any attempt to eliminate the most linearly correlated features is likely to discard many important features and curtail the power of the remaining feature set. From our experiments we found that feature sets whose covariance matrices are sufficiently independent to obtain their inverses perform no better than random feature sets of the same size. Therefore, we conclude, that the contribution of linearly dependent features within non-linear classifiers, can be significant as was shown in Chapter 4.

When we discussed the features we use for our spatial analysis we introduced the Entropy Distance measure $(D_H(Y, X))$, which is a linear combination of the Joint Entropy $(H(X, Y))$ of the two frames and the Mutual Information $(H(Y; X))$.

$$D_H(Y, X) = H(X, Y) - H(Y; X)$$

There is nothing any classifier can find out from $D_H(X, Y)$ that it does not already know from $H(X, Y)$ and $H(Y; X)$. Adding the Entropy Distance to the set of features will prevent us from using any covariance-based classifier. In fact, the feature set is so redundant that we cannot use any covariance-based methods to estimate the redundancy/relevance of the individual features. But, why did we add the Entropy Distance to our feature set in the first place? The answer is, "because we can", provided that we don't use a covariance-based classifier to classify our data. Although, we can not assume that the Entropy Distance captures all the information of the Joint Entropy and the Mutual Information, it is possible that it captures all the *relevant information* we need about them. If this were true – and we never know for incomplete data – then we could reduce the dimensionality of our representation of the world (i.e., the video frames) by using the Entropy Distance function alone instead of its **two** components the Joint Entropy and the Mutual Information. This again shows that limiting our search for good measures to those that are linearly uncorrelated to any of the measures we already know about may actually bloat the set of measures. This is especially true once our set of measures becomes larger than, say, 10, or so. Now there are so many linear combinations between pairs, triples, quadruples, etc. of measures that it becomes totally impossible to guarantee optimality of the set. These issues are all strong support for non-covariance-based classifiers, such as neural networks.

1.3.   Linear and Quadratic Classifiers

Many of the available classifiers make implicit assumptions that the different classes/groups have similar covariance matrices or are normally distributed. If these assumptions are violated many of the derived results become invalid, and any "optimality" claimed by the classifiers is no longer guaranteed. Section 1 of Chapter 4 showed that classical linear methods are inferior to non-linear methods, such as a back-propagation neural network.

1.4. Random Feature Sets

For k features there are 2^k possible subsets of features and transitivity cannot be used to build *optimal* sets of size $N+1$ from optimal sets of size N. Likewise we cannot use transitivity to obtain optimal subset of size $N-1$ by eliminating the least important feature from optimal sets of size $N-1$. However, we have shown how to find *good* feature sets by using a greedy strategy on good random collections of features (see Section 2.4 of Chapter 4)

The best 13-feature set found combining good random sets with a greedy strategy, spans all but one feature extraction method and out-performs all the methods in isolation (even the combined set of 28 GLCM based features). Generally, random collections of features from different methods out-perform equally sized feature sets of one type.

1.5. Back-Propagation Neural Networks

The use of a back-propagation network offers a simple solution to the laborious task of finding a good combination of the available features, thus solving our fusion problem. We have shown that feature sets like the one presented have sufficient expressive power to allow good generalization from only a few training images. An analytical approach, on the other hand, is difficult to conduct since the interactions even between modest numbers of dependent features are complex.

2. Object Recognition

Our feature relevance analysis supports our approach of gathering rich image descriptions based on a number of fundamentally different measures. The neural network classifier performed best when information from 6 of the 7 feature extraction methods was available. The methods include measures based on the Fourier transform, Gabor filter banks, the fractal dimension, color, entropy, and gray-level co-occurrence matrices. We had expected a stronger contribution from the steerable filter based features. We hoped that the associated line and step-edge analysis would aid the detection of branches, forks, junctions, and human made structures (with bar- and step-edges at a few distinct orientations). In particular the detection of deciduous trees in winter, or the detection of human-made structures in the landing, and rocket-launch events should have benefited from this analysis. However, we found that the responses of these filters are too dependent on mean gray-level and contrast changes in images. A low contrast image of a bar-edge with many other weak and scattered bar- or step-edges in the same image patch often results in too little support for the bar-filter to justify labeling the patch as a bar- edge. This is often the case even if the only consistent structure in the image is bar-like as shown on the right of Figure 5.1. Labeling the large branch in this image as a bar-edge is

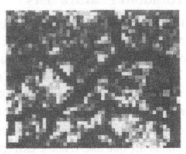

Figure 5.1. The right image is a magnified version of the central region of the left image. Labeling the large branch in the right image as a bar-edge is complicated by the presence of the smaller branches and leaves in front and behind it. Most edge- or bar-detectors are optimally designed for clean step- and bar-edges, however, they perform poorly for cluttered images.

difficult because of the smaller branches and leaves in front and behind it. It is likely that a steerable filter will align itself well along the branch, however, the strength of the response of the steered filter will be much weaker than the response of both the bar- as well as the step-edge filter on an isolated bar (or even a step-edge) with a high contrast. Some normalization of the response has to occur before steering the filters will be reliable and indicative enough for detection of branches, forks and intersections.

Another issue is that to exploit steerable filters more appropriately some non-local techniques might be necessary that establishes the presence of long line segments of significance.

The use of a back-propagation network offers a simple solution to the laborious task of finding a good combination of the features. We have shown that feature sets like the one presented have sufficient expressive power to allow good generalization from only a few training images. Since the back-propagation algorithm is well-understood and analyzed we have shown that it is possible to determine the usefulness of a specific feature if we had to reduce the amount of features used to obtain a subset. The neural network approach offers a synthetic solution to the sensor fusion problem that is concerned with combinations of (possibly dependent) features for the purpose of classification and/or recognition. An analytical approach on the other hand would be difficult to conduct since the interactions even between modest numbers of dependent features are complex.

3.   Event Detection

Our experimental results have verified the effectiveness of the presented framework. The framework decomposes the task of extracting semantic events into three stages where visual information is analyzed and abstracted. The first stage extracts low-level features and is entirely domain-independent. The second stage analyzes the extracted low-level features and generates intermediate-level descriptors, some of which may be domain-specific. In this stage, shots are summarized in terms of both domain-independent and domain-specific descriptors. To generate the shot summaries, regions of interest are detected, verified and tracked. The third and final stage is domain-specific. Rules are deduced from specific domains and an inference model is built based on the established rules.

In other words, each lower stage encapsulates low-level visual processing from the higher stages. Therefore, the processes in the higher stages can be stable and relatively independent of any potential detail changes in the lower level modules. To detect different events, the expected changes are (a) the addition of descriptors in the second stage and (b) the design of a new set of rules in the third stage. The presented framework also provides several reusable algorithmic components. In fact, the extracted low-level texture and color features are domain independent and many objects involved in events carry certain texture and color signatures. The neural network used for image region classification can be easily re-configured or extended to handle other types of objects [50]. The robust statistical estimation based object tracking method has already been used in different applications, and its robustness and simplicity are verified in experiments repeatedly [97].

The presented framework detects events by detecting spatio-temporal phenomena, which are physically associated with the event in nature. More precisely, the physical phenomenon, which we attempt to capture is, the combination of the presence of objects in space and their movement patterns in time. This is in contrast to many existing event detection methods, which detect events by detecting artificial post-production editing patterns or other artifacts. The drawbacks of detecting specific editing patterns or other artifacts are that those patterns are often content provider dependent and it is difficult, if not impossible, to modify the detection methods and apply them to the detection of other events. Our framework solves practical problems, and the solutions are needed in the real world. In many of the video tapes which we obtained, the speech from the audio track and the text from the close-caption are only loosely correlated with the visual footage. It is therefore unlikely that the event segments may be accurately located by analyzing the audio track and close-caption alone. Given the existing video data, a visual-information-based detection algorithm is needed to locate the event segments; otherwise, manual annotation is required.

3.1. Accuracy, Robustness, and Scalability

Much research in computer vision has focused on the precise extraction of object and motion information from images and image sequences. In many cases though such precision is not needed. We have shown that hunts, landings, and rocket launches can be detected without the determination of shape and even identity of the hunting, landing, or launched object. In many cases the detection of events may provide essential functional, causal, and contextual information about objects, without which their detection or recognition may be difficult or even impossible. For instance, object recognition is sometimes seen as having three parts [44, 46]:

Selection : Given a set of data features, extract (possibly overlapping) subsets that ar likely to have come from a single object.

Indexing : Given a library of possible objects, select a subset that are likely to be in the scene, perhaps as a function of the selected data subsets.

Correspondence : For each subset from the selection step, and for each corresponding object from the indexing step, determine if a match can be found between a subset of the data features and a subset of the model features, consistent with a rigid transformation of the object.

The selection and indexing problems are exponential: there are 2^n ways to choose subsets of a set of size n. Allowing spurious data, due to clutter, occlusion, specular reflections, shadows, etc. the correspondence problem is exponential for every pair of selected features and indexed objects. Grouping, selecting, and indexing features and objects are therefore vital to facilitate the search for an appropriate object model for the objects in an image [44, 46].

In other cases it may be (practically or theoretically) impossible to recognize an object by its

shape : e.g., trees, clouds, rocks, chairs, U.F.O.s, etc. have no standardized shape, or new shapes are created continuously, e.g., cars, planes, houses, etc.,

color : e.g., most human made objects can be made in any color,

texture : e.g., lighting conditions may impose texture on objects that are not indicative of the objects, etc. Imaging them in motion may impose/distort/remove texture.

The applications presented in this book were shown to be insensitive to a wide range of color and texture changes of the detected objects and do not require the specification of shape for the objects involved. If external segmented object information is available, the presented framework can integrate it, by bypassing the object recognition components. If internal or external information suggests that specific objects are present in the input video then the object recognition component can search for it explicitly. If no object information is available this framework may provide essential cues that enable their robust recognition given information about the events in which they are involved.

Likewise, the framework could benefit from more precise and more detailed motion information. The global motion model is coarse and additional motion information for the various object regions in the image would simplify the design of event detectors. Particularly the use of a background mosaic could simplify and robustify foreground segmentation.

4. Building Mosaics from Video

The results presented in the previous chapter, were intended to demonstrate that robust visual event detection can be achieved by combining relatively crude, low-level spatial image measures with simple motion measures, a simple foreground/background model, and simple and easily designed event models.

Here we explore an alternative to the presented motion estimation method. A common approach that often achieves better segmentation results is based on background-differencing (or background subtraction). This approach often enables the segmentation of the entire foreground object from the background, when frame-differencing methods can only detect the leading and trailing edges of a foreground object. Figure 4.15 clearly shows this observation. Only the leading and trailing edges of the moving object are sufficiently different to stand out as foreground. We managed to get around this problem by assuming that there is a single dominant motion-blob in every frame, whose location we could determine even from the leading and trailing edges.

To avoid the problems associated with frame-to-frame segmentation of foreground we need a smarter approach that combines the information from multiple frames to realize that the entire cheetah is a foreground object. One such segmentation approach is based on background differencing. Background differencing labels everything as foreground that is not part of the background mosaic. If, as the camera is moving we manage to identify all the objects whose motion is not explained by the camera motion, and we build a mosaic of the true background regions we can use that background in a second pass over the relevant part of the video to segment the foreground from the background at every frame. With the exception of boundary conditions foreground objects will be segmented entirely, not just their leading and trailing edge, as is usual with frame-to-frame differencing based segmentation methods. See for instance, [62, 109] for details on how to assemble video frames into mosaics.

The top of Figure 5.2 shows a background mosaic constructed from hundreds of video frames. The first frame is shown below the mosaic. New images in the sequence are registered and then merged with the mosaic. Merging is often done by blending the contributions of the mosaic and the new image. To limit the effects of lens distortion blending is often done using a weighting function that emphasizes the pixels at the center of a frame and de-emphasizes contributions from peripheral pixels. The images and the algorithms used to construct the mosaics are courtesy of DiamondBack Vision, Inc. [30].

Figure 5.2. A mosaic extracted from a video sequence

Figure 5.3 shows how these background models can be used to segment foreground. Again, The images, segmentations, and the mosaics are courtesy of DiamondBack Vision, Inc. Figure 5.3 (b) shows a frame from the video sequence that was used to construct the mosaic in (a). Figure 5.3 (c) shows the unwarped background for a frame from the center of the mosaic. Figure 5.3 (d) shows the foreground found by background subtraction for the same frame (anything except for the checker board pattern represents foreground)

To estimate camera motion using an affine model we need to determine at least 3 correspondences between frames. To estimate camera motion using a projective model we need to determine at least 4 correspondences between frames. To make these methods robust to noise many more than the theoretically sufficient 3 or 4 correspondences are needed. In fact, many tens or even hundreds of correspondences may be necessary to obtain good motion estimates. The correspondences are used with least squares methods to estimate the 6 affine or 7 projective parameters, respectively. The correspondences are often determined by using the output of a corner detector run on both frames. Corner detectors attempt to find points in the image that are likely to have unique spatial correspondences. Points on a line or step-edge, for instance, are good to constrain motion estimates in the direction perpendicular to the line or step-edge, but often many other points along the line or edge have very similar properties and may thus lead to bad motion estimates. The uncertainty associated with such points is commonly referred to as the "aperture" problem. The features we already extracted for our spatial analysis can be used to further disambiguate between correspondences detected by the corner detector.

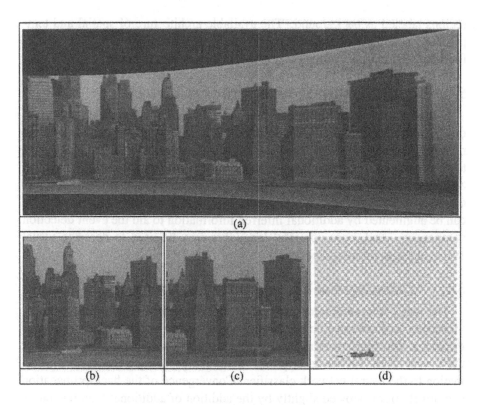

Figure 5.3. Background differencing using a mosaic

The following issues often complicate building good background models:

- Aperture problems are often caused by motion blur or a general lack of texture, thus reducing the number of correspondences that can be used to estimate camera motion

- Violations of the camera motion model, such as parallax, mixtures of camera rotations and translations in an image region, lens distortion, etc. prevent the model from estimating motions correctly

- Lighting and the camera's Automatic Gain Control (AGC) violate the assumption that the only thing that happened between the two snap-shots of the **static** world is the true camera motion

- Finally, the static world assumption may itself be wrong. The video frames may contain any number of independently moving objects.

The last issue usually makes it necessary to segment video frames into foreground and background regions, and to build background models only using regions labeled as background. The available architecture of spatial and temporal components is sufficient to achieve such segmentations. See [103, 30, 62] for details on mosaic construction and background differencing.

5.   Improving the Modules of the Framework

In this section we cast some light on how the functionality of the modules of our framework might be improved. The flat hierarchy of abstractions and the modular structure allow for significant changes without necessitating changes to the framework. Better motion and object segmentation can be achieved by improving the respective components of the framework, better classifiers can be used to better fuse the color and texture information. Shot summaries can be augmented by additional inferred information to aid the event detection component. And more powerful languages may be used to describe visual events at the event inference level.

5.1.   Color, Spatial Texture, and Spatio-temporal Texture

The spatio-temporal color and texture analysis module can be improved by increasing its expressive power through the inclusion of new color and texture measures. Our framework naturally utilizes additional information provided at the lowest level. Richer descriptions of the image will provide further information to the neural network classifier. Convergence of the back-propagation neural network is slowed slightly by the addition of additional features, but at the same time, the performance of the neural network monotonously increases with the feature set size.

5.2.   Motion Estimation

The motion analysis module can make use of more elaborate algorithms to obtain more accurate, robust, and detailed motion estimates. So far we

1 estimate the dominant motion in video frames,

2 compensate for this motion between consecutive frames,

3 and locate a single motion-blob corresponding to the region with the largest residual motion errors after motion compensation.

More powerful motion models can be used to extract more precise foreground and background motion information. Likewise, background differencing may be used to improve motion-based segmentation. A large number of schemes exist that estimate the dominant motion in video sequences, such as [103, 62]. Those methods might improve the motion estimation results and thus the following motion compensation step. The last step might benefit from a more flexible motion estimation scheme that partitions video frames into more than a dominant and an independent motion region. However, allowing more than a single independent motion-blob requires the higher-level modules to be able to process this information. Disjoint motion-blobs might correspond to a single, partially occluded object. Likewise a single motion-blob might consist of two overlapping, moving objects. In general this problem is known as the segmentation problem. Often segmentation is impossible without prior recognition of the object, and vice versa, and object recognition is impossible without prior segmentation. The use and combination of motion, color and texture information can aid the segmentation process but cannot solve it without error.

5.3.   Object Classification

Better object classification is likely to be achievable if motion information is passed to it. The classifier is not given explicit motion information at each image location. A background differencing scheme combined with an affine or projective motion model can provide valuable new information to the classifier. Many objects are not normally associated with any *significant* ego-motion, such as buildings, trees, soil, or mountains. Other objects are associated with characteristic motions, such as the rotational motions of wheels, or drums, the up-and-down motions of seesaws, upward-motions of rockets, translational motions of trains or cars, or the walking, jumping, crawling, motions of animals and humans. In the events we presented in this book we have shown how qualitative motion can be used at a higher-level to extract event information from video sequences. It is likely that the same information injected at a lower level can aid the object classifier to achieve better recognition rates.

5.4.   Shot Boundary Detection

The shot boundary detection module can be optimized to account for more complex dissolves and fades between shots that may remain undetected by the current module. The classifier may be improved by replacing it with another type of classifier, by providing further information to it, such as motion information, or feedback from higher-level modules, such as feedback from its own output on earlier frames, feedback from the motion-blob verification process, from the shot summarizer or even from the event detector.

5.5.   Feedback

Since the flow of information is strictly bottom-up in the presented framework it is generally possible to augment the information available to the various modules by including top-down information. Just as this might help the classifier as was just pointed out, the motion-blob verifier might use information from the shot summarizer and the event detector to provide a framework within which the task of verifying motion-blobs might be greatly simplified. The focusing of the attention of a module might in itself be useful information with which the video sequence should be augmented or which should be forwarded to the user/viewer to improve the transparency of the event detector.

5.6. Shot summarization

The shot summarizer is the most important bottleneck in the framework since it combines information from all lower level modules for the presentation to the event detector that has to make a hard decision as to the presence or absence of a particular event. While the event detector is largely a logical representation of actions on objects over time, it is the shot summarizer's task to provide the predicates used by the event detector module, and to assign them values that correspond to those a human observer would assign. A richer spatio-temporal "vocabulary" may be necessary to enable wider coverage of different domains and events and to make sure that new events can be described in terms of existing descriptors.

5.7. Learning Event Models

It took evolution several billion years to produce good visual sensors. It takes us a good part of our first year in school to learn to recognize the 26 characters of the alphabet. But a cat only needs to be shown how to get out of the house once, and it will never forget. These are strong indications that

1 Bottom level processing is crucial to provide robust and versatile high-level processing. The fact that most of the low-level capabilities of living beings are hard-wired into their genes indicates that nature has no patience for a learning system to achieve the same in "soft-ware".

2 Object recognition can be learned but requires multiple passes through the data set (see how long it takes you to learn the Korean alphabet, of 24 characters).

3 The detection and recognition of sequences is often trivial. You don't forget how to get to your fridge, not even on the first day in your new apartment.

Thus it is not surprising that describing the low-level features takes significantly longer than to describe the algorithm of our classifier. In turn, describing the event inference module in terms of the accumulated information is trivial. A small sketch of the intuitive relationships between the various objects and event phases is all that is necessary to capture the event model. We are optimistic that learning techniques can be used to automatically infer or at least optimize the event model.

5.8.   Tracking

The presented rich image descriptions might be a powerful tool to robustly track arbitrary objects. Activated by the detection of a motion-blob a range of statistics of an independently moving image region can be gathered and used to keep track of the corresponding object as the object moves in its environment. This approach could eliminate the need for parameters for general object trackers. It is also likely that no prior knowledge of the objects is needed to enable their tracking, since the rich image descriptors serve as a signature for the object as long as it does not change its appearance (through out-of-plane rotations, morphing, articulated part movements, etc.).

5.9.   Kullback-Leibler Divergence for Correspondence Matching

A common approach to the determination of correspondences in related images/frames is to use correlation approaches. These approaches use the appearance of an object in one image (an image patch) to locate the same object in another image/frame. While this approach can be made illumination invariant by normalizing the image patches before matching them, it is difficult to use this approach when non-linear changes exist between the image patches. For instance, partial shadowing, or deformation of the object will deteriorate the performance of correspondence matches based on the correlation between the image patches.

The relative entropy measure, described in Section 1.7 of Chapter 2, could offer a solution for these problems, since patches are not compared directly but rather indirectly, via their histograms. This renders this approach somewhat invariant to partial shadowing, lighting changes, camera parameters, or deformations and may improve the performance of the matcher in such cases.

It would be interesting to see where entropy, or other color/texture-based, correlation can improve on intensity based correlation and if general observations can be made that indicate when correlation based trackers should be used and when relative entropy trackers should be used. Another issue would be to investigate whether we could automatically switch between both approaches depending on the patches themselves, the history of the tracked object, or observed changes in lighting conditions or deformations in the object, as detected by other means.

5.10.   Temporal Textures

It may be possible to recognize objects merely by their spatio-temporal appearance, entirely without instantaneous spatial information. The mutual information should be low for leaves in the wind, since consecutive frames are likely to be only weakly dependent. Each pixel may have five states: occluded, non-occluded direct light, non-occluded indirect light, specular reflection, and background. Most other objects, natural or human-made are unlikely to share these properties.

For waves on a water surface, each pixel may have six states: transparent, reflection, specular reflection, polarized reflection, direct light, and indirect light. For human made objects, each pixel may have two states: direct light and indirect light.

These observations should manifest themselves in the described entropy spatio-temporal measurements, and should permit the detection of a number of objects from video sequences, based only on their spatio-temporal appearance.

6.   Applications

We have motivated the use of rich image descriptions to aid classification and event detection in a flat hierarchy of abstractions. Next we discussed the architecture of the framework and its components, before we showed some results for the object classifier and the event detector. Finally, we discussed alternatives, variations, and possible improvements for the framework. But are the three events shown here the only ones that can be modeled in this way? This section will list a number of events that we believe can be modeled using the same framework. All of the events require modifications to the event model. Many may also depend on the extraction of different shot statistics and shot properties. Some may need the neural network classifier to be trained on a new object class. But, we believe that none depend on additional low-level information, or changes to the architecture of the framework. Table 5.1 shows a list of sample events and actions that might benefit from the presented spatio-temporal video descriptors:

Table 5.1.   Further events that can be modeled using the presented framework.

Traffic	Surfing	Snow-boarding
Climbing	Paragliding	Wind-surfing
Surf	Ski-jumping	Pole-vaulting
Fire	Flags in the wind	Leaves in wind
Rain	Waterfalls	Whitewater rafting
Crowds	People eating	Long-jumping
Waterfalls	People smoking	People on the phone
Smoke/Fog	Hurdle races	People in a meeting

The next section details possible stages and strategies for some of these events.

6.1. Recipes for Selected Applications

Traffic/flow:

- static camera,
- motion only in fixed region of image,
- objects constantly enter and exit the image,
- objects have similar velocity, texture and color (car traffic might differ in color).

Ski jump:

- start : static camera, possibly moving ski jumper, white below motion-blob of ski jumper,
- descend : moving camera, increasing speed, tracking ski jumper, white below tracked ski jumper,
- jump : very fast motion strong downward component, no constant white areas in field of view, shape of motion-blob like horizontally flat ellipsoid,
- landing : white becomes visible again and increases in area within frame, shape of motion-blob changes to upside-down "T",
- slowdown: motion of ski jumper decreases, ski jumper occupies larger area in frame,
- postjump: scoreboard, numbers, crowds.

Long jump:

- start : no motion,
- sprint : increasing motion,
- jump : change of ground texture/color, motion periodicity stops,
- landing: abrupt termination of motion, possibly change of camera, color and texture of sand,
- postjump: scoreboard, numbers, crowds.

High jump:

- start : no motion,
- sprint : increasing motion,
- jump : change of ground texture/color, motion periodicity stops, sky visible, two long lines visible, negation of vertical motion component,
- landing: abrupt termination of motion, possibly change of camera, color and texture of landing pad,

- postjump: scoreboard, numbers, crowds.

Aircraft/Birds taking off:

- start : slow or no motion,
- acceleration: increasing motion magnitude, motion-blob orientation changes by raising front, camera tracking object,
- lift off : background change, motion upward, sky visible, motion too difficult to measure or near zero, size of motion/texture blob decreasing.

Waterfall: Still camera, fixed white area in image that shows motion, but doesn't move, clouds of steam at bottom of waterfall, blue water below waterfall, possibly with some constant motion in a direction different from the direction of the white, falling water, rock or trees/vegetation visible on either side of the waterfall.

Surf: Still camera, horizon, no motion above horizon, possibly no motion in foreground on beach, band of periodic motion in between with lots of white/green/blue in horizontally elongated patterns.

Flag in the wind: Still camera, sky, periodic appearance and disappearance of the flag, similar motion, texture, and color statistics in fixed area of image, usually doesn't extend from edge to edge (of frame).

Appendix A

To prove the *Data Processing Theorem*, we need to show that

$$H(W; R) - H(W; D) \leq 0$$

We do this as follows:

$$
\begin{aligned}
H(W; R) - H(W; D) &\overset{def}{=} H(W) - H(W|R) - H(W) + H(W|D) \\
&= -H(W|R) + H(W|D) \\
&= -\sum_{w,r} P(w,r) \log \frac{1}{P(w|r)} \\
&\quad + \sum_{w,d} P(w,d) \log \frac{1}{P(w|d)} \\
&= \sum_{w,r} P(w,r) \log P(w|r) \\
&\quad + \sum_{w,d} P(w,d) \log \frac{1}{P(w|d)} \\
&= \sum_{w,d,r} P(w,d,r) \log \frac{P(w|r)}{P(w|d)} \\
&= \sum_{w,d,r} P(d,r) P(w|d,r) \log \frac{P(w|r)}{P(w|d)}.
\end{aligned}
$$

The last step follows from the definition of conditional probabilities,
$P(w|d,r) = \frac{P(w,d,r)}{P(d,r)}$. Since $P(w,d,r) = P(w)P(d|w)P(r|d)$, we further
have

$$P(w|d,r) = \frac{P(w)P(d|w)P(r|d)}{P(d,r)}$$

$$= \frac{P(w)P(d,w)P(r,d)}{P(d,r)P(w)P(d)}$$

$$= \frac{P(d,w)}{P(d)}$$

$$= P(w|d).$$

Plugging this result into equation (5.0) we obtain

$$H(W;R) - H(W;D) = \sum_{w,d,r} P(d,r)P(w|d) \log \frac{P(w|r)}{P(w|d)}$$

$$= \sum_{d,r} P(d,r) \left[\sum_{w} P(w|d) \log \frac{P(w|r)}{P(w|d)} \right].$$

Since $0 \le \frac{P(w|r)}{P(w|d)} \le 1$, we have $log \frac{P(w|r)}{P(w|d)} \le 0$. Therefore, the quantity in brackets is minus a relative entropy, and $H(W;R) - H(W;D) \le 0$.

References

[1] A. Akutsu and Y. Tonomura, "Video Tomography: An Efficient Method for Camerawork Extraction and Motion Analysis," in the proceedings of *ACM Multimedia*, ACM, 1994.

[2] T.D. Alter and D. W. Jacobs, "Error Propagation in Full 3D-from 2D Object Recognition", *Computer Vision and Pattern Recognition 1994*, pp. 892-899, 1994.

[3] F. Arman, R. Depommier, A. Hsu, and M.-Y. Chiu, "Content-based Browsing of Video Sequences," in the proceedings of *ACM Multimedia*, pp. 97-103, 1994.

[4] V. Athitsos, M.J. Swain, and C. Frankel, "Distinguishing Photographs and Graphics on the World Wide Web," in the proceedings of *IEEE Workshop on Content-based Access of Image and Video Libraries*, in conjunction with CVPR'97, pp. 10-17, 1997.

[5] N. Ayache and O.D. Faugeras, "Artificial Vision for Mobile Robots: Stereo Vision and Multisensory Perception,", MIT Press, 1991.

[6] D. Ayers and M. Shah, "Monitoring Human Behavior in an Office Environment," in the proceedings of *Workshop on Interpretation of Visual Motion*, in conjunction with CVPR'98, 1998.

[7] A. Azerbayejani, *et. al.*, "Real-Time 3D Tracking of the Human Body", *MIT Media Lab, Perceptual Computing Section, TR No. 374*, 1996.

[8] S. Barnard, "A Stochastic Approach to Stereo Vision", in *Readings in Computer Vision: Issues, Problems, Principles and Paradigms*, pp. 21-25, 1987.

[9] P. Belhumeur and D. Kriegman, "What Is the Set of Images of an Object Under All Possible Illumination Conditions?," Int. Journal of Computer Vision, 28(3), pp. 245-260, 1998.

[10] P. Belhumeur and G. Hager, "Efficient Region Tracking with Parametric Models of Geometry and Illumination," IEEE Trans. PAMI, 20(10), pp. 1025-1039, October 1998.

[11] M. J. Black and Y. Yacoob, "Tracking and Recognizing Rigid and Non-Rigid Facial Motion using Local Parametric Models of Image Motion," in the proceedings of the *International Conference on Computer Vision*, 1995.

[12] Sing-Tze Bow "Pattern Recognition and Image Preprocessing," Dekker, 1992.

[13] S. Belongie, C. Carson, H. Greenspan, J. Malik, "Color- and Texture-Based Image Segmentation Using EM and its Application to Content-Based Image Retrieval", in IEEE Workshop on *Content based Access of Image and Video Databases*, in conjunction with ICCV'98, 1998.

[14] K. Bowyer, M. Sallam, D. Eggert, and J. Stewman, "Computing The Generalized Aspect Graph For Objects With Moving Parts", in *IEEE Transaction on Pattern Analysis and Machine Intelligence*, 15:605-610, 1993.

[15] R.A. Brooks, "A Robust Layered Control System for a Mobile Robot," A.I. Memo 864, MIT, 1985.

[16] O.I. Camps, "Towards a Robust Physics Based Object Recognition System," *Lecture Notes in Computer Science (994): Object Representation in Computer Vision*, Springer-Verlag, 1995.

[17] J.F. Canny, "Finding Edges and lines in Images," Masters Thesis, Massachusetts Institute of Technology, 1983.

[18] C. Carson, S. Belongie, H. Greenspan, and J. Malik, "Color- and Texture-Based Image Segmentation using EM and its Application to Image Querying and Classification," submitted to IEEE for possible publication, 1998.

[19] S.-F. Chang, W. Chen, H.J. Meng, H. Sundaram, and D. Zhong, "A Fully Automated Content Based Video Search Engine Supporting Spatio-Temporal Queries," in *IEEE Transactions on Circuits and Systems for Video Technology*, 1998.

[20] B.B. Chaudhuri, N. Sarkar, and P. Kundu, "Improved Fractal Geometry Based Texture Segmentation Technique," in *IEE Proceedings*, part E, vol. 140, pp. 233-241, 1993.

[21] M.B. Clowes, "On Seeing Things", in *Artificial Intelligence*, 2, No. 1, pp. 76-116, 1971.

[22] R.W. Conners, C.A. Harlow, "A Theoretical Comparison of Texture Algorithms," in *IEEE Transactions on Pattern Analysis and Machine Intelligence*, vol. 2, no 3, pp. 204-222, 1980.

[23] T.N. Cornsweet, "Visual Perception", Academic Press, New York, 1970.

[24] J.D. Courtney, "Automatic Video Indexing via Object Motion Analysis," *Pattern Recognition*, vol. 30, no. 4, pp. 607-626, 1997.

[25] T. Cover and J. Thomas, "Elements of Information Theory," *Wiley & Sons*, New York, 1991.

[26] G. Cybenko, "Approximation by Superposition of Sigmoidal Function," *Mathematics of Control, Signals, and Systems*, Chapter 2, pp. 303-314, 1989.

[27] M. Davis, "Media Streams: An Iconic Visual Language for Video Annotation" in the proceedings of *1993 IEEE Symposium on Visual Languages in Bergen, Norway*, IEEE Computer Society Press, pp. 196-202, 1993.

[28] J. Davis and M. Shah, "Visual Gesture Recognition", in IEE proceedings of *Vision, Image and Signal Processing*, Vol. 141, No. 2, pp. 101-106, 1994.

[29] Y.F. Day, A. Khokhar, S. Dagtas, and A. Ghafoor, "Spatio-Temporal Modeling of Video Data for On-line Query Processing," in the proceedings of *IEEE International Conference on Multimedia Computer Systems*, 1995.

[30] DiamondBack Vision, Inc, "Video Segmentation using Statistical Pixel Modeling," US Patent filed, 2001.

[31] A. Del Bimbo, E. Vicario, D. Zingoni, "A Spatial Logic for Symbolic Description of Image Contents," in the *Journal of Visual Languages and Computing*, vol. 5, pp. 267-286, 1994.

[32] A.P. Dempster, N.M. Laird, and D.B. Rubin, "Maximum Likelihood from Incomplete Data via theEM algorithm", in *Journal of the Royal Statistical Society, Series B, 39(1): 1-38*, 1977.

[33] N. Dimitrova and F. Golshani, "Motion Recovery for Video Content Classification," in *ACM Transactions on Information Systems*, vol. 13, no 4, pp 408-439, 1995.

[34] P. England, R.B. Allen, M. Sullivan, and A. Heybey, "I/Browse: The Bellcore Video Library Toolkit," in SPIE proceedings on *Storage and Retrieval for Image and Video Databases*, pp. 254-264, 1996.

[35] T.Kato, T. Kurita, and H. Shimogaki, "Intelligent Visual Interaction with Image Database Systems - Toward the Multimedia Personal Interface," in the *Journal of Information Processing,*, Vol. 14, No. 2, 1991, pp. 134-143.

[36] S. Fahlman, "Faster-Learning Variations on Back-Propagation: An Empirical Study," in the proceedings of the *Connectionist Models Summer School*, Morgan Kaufmann, 1988.

[37] O. Faugeras, *Three-Dimensional Computer Vision: A Geometric Viewpoint*, MIT Press, 1993.

[38] R.A. Fisher, "The use of multiple measurements in taxonomic problems," in *Ann. Eugenics 7 (part 2) pp. 179-188*, 1936.

[39] M.A. Fischler "Robotic Vision: Sketching Natural Scenes", in the proceedings of the *1996 ARPA IU Workshop* , 1996.

[40] I.Fogel and D.Sagi, "Gabor Filters as Texture Discriminator," in the *Journal of Biological Cybernetics*, vol. 61, pp. 103-113, 1989.

[41] W.T. Freeman and E.H. Adelson, "The Design and Use of Steerable Filters," in *IEEE Transactions on Pattern Analysis and Machine Intelligence*, Vol. 13, pp. 891-906, 1991.

[42] D. Gabor, "Theory of communication," in *Journal of the IEE*, vol. 93, pp. 429-457, 1946.

[43] R. Gonzalez, "Digital Image Processing," published by Addison-Wesley

[44] W.E.L. Grimson, "The Combinatorics of Heuristic Search Termination for Object Recognition in cluttered environments," A.I. Memo No. 1111, MIT, Cambridge MA, May 1990.

[45] W.E.L. Grimson, "Object Recognition by Computer," MIT Press, Cambridge, MA, 1990.

[46] W.E.L. Grimson, "The Effect of Indexing on the Complexity of Object Recognition," A.I. Memo No. 1226, MIT, Cambridge MA, April 1990.

[47] A. Guzman, "Decomposition of a Visual Scene into Three-Dimensional Bodies," in the proceedings of the *FJCC*, 1968.

[48] G.D. Hager and P.N. Belhumeur, "Real-Time Tracking of Image Regions with Changes in Geometry and Illumination," in the proceedings of the *Conference on Vision and Pattern Recognition*, 1996.

[49] D.J. Hand, "Construction and Assessment of Classification Rules," *Wiley*, 1997.

[50] N. Haering, Z. Myles, and N. da Vitoria Lobo, "Features and Classification Methods to Locate Deciduous Trees in Images", in Journal on Computer Vision and Image Understanding, pp. 133-149, 1999.

[51] N. Haering, R.J. Qian, and M.I. Sezan, "Detecting Hunts in Wildlife Videos," in the proceedings of the *International Conference on Multimedia Computing and Systems*, 1999.

[52] R.M. Haralick, K. Shanmugam, and I. Dinstein, "Textural Features for Image Classification," in *IEEE Transactions on Systems Man and Cybernetics*, vol. 3, no 6, pp. 610-621, 1973.

[53] D. Panjwani and G. Healey, "Markov Random Field Models for Unsupervised Segmentation of Textured Color Images," in *IEEE Transactions on Pattern Analysis and Machine Intelligence*, 17(10):939-954, October, 1995.

[54] M. Hoetter, "Differential Estimation of the Global Motion Parameters Zoom and Pan," in *Signal Processing*, (16), pp. 249-265, 1989.

[55] B.K.P. Horn, "Shape from Shading: A Method for Obtaining the Shape of A Smooth, Opaque Object from One View", PhD thesis, Massachusetts Institute of Technology, 1970.

[56] B.K.P. Horn, "Height and Gradient from Shading", in *International Journal of Computer Vision*, 5:37-75, 1990.

[57] M.R. Naphade, T. Kristjansson, B. Frey, and T.S. Huang, "Probabilistic Multimedia Objects (Multijects): A Novel Approach to Video Indexing and Retrieval in Multimedia Systmes", in the proceedings of the *International Conference on Image Processing*, 1998.

[58] D.A. Huffman, "Impossible Objects as Nonsense Sentences," in *Machine Intelligence*, **6**, pp. 295-323, 1971.

[59] M.S. Lew, K. Lempinen, N. Huijsmans, "Webcrawling Using Sketches", in the proceedings of *Visual'97*, 1997.

[60] S.S. Intille, "Tracking Using a Local Closed-World Assumption: Tracking in the Football Domain," Master Thesis, Massachusetts Institute of Technology, Media Lab, 1994.

[61] M. Irani, P. Anandan, and S. Hsu, "Mosaic Based Representations of Video Sequences and Their Applications", Proc. 5th International Conference on Computer Vision, pp. 605-611, 1995.

[62] M. Irani and P. Anandan, "A unified approach to moving object detection in 2D and 3D scenes," in IEEE Transactions on *Pattern Analysis and Machine Intelligence*, pp. 707-718, 1997.

[63] G. Iyengar and A. Lippman, "Models for Automatic Classification of Video Sequences", *SPIE Proc. Storage and Retrieval for Image and Video Databases*, pp. 216-227, 1997.

[64] D.W. Jacobs, "Recognizing 3-D Objects Using 2-D Images," PhD Thesis, Massachusetts Institute of Technology, 1992.

[65] R. Jain, R. Kasturi and B. Schunck, "Machine Vision," *McGraw Hill*, 1995.

[66] P. Kanerva, "Sparse Distributed Memory" *MIT Press*, 1990.

[67] T. Kawashima, K. Tateyama, T. Iijima, and Y. Aoki, "Indexing of Baseball Telcast for Content-based Video Retrieval," in the proceedings of the *International Conference on Image Processing*, pp. 871-875, 1998.

[68] J.M. Keller and S. Chen, "Texture Description and Segmentation through Fractal Geometry," in *Journal of Computer Vision, Graphics and Image Processing*, vol. 45, pp. 150-166, 1989.

[69] S. Khan and M. Shah, "Tracking People in Presence of Occlusion", in *Asian Conference on Computer Vision*, 2000.

[70] J. Krumm, "Space Frequency Shape Inference and Segmentation of 3D Surfaces", PhD Thesis, Carnegie Mellon University, 1993.

[71] S.-Y. Lee, M.-K. Shan, W.-P. Yang, "Similarity Retrieval of Iconic Image Database," in *Pattern Recognition*, Vol. 22, No. 6, pp 675-682, 1989.

[72] Y. Bengio, Y. LeCun, D. Henderson, "Globally Trained Handwritten Word Recognizer using Spatial Represention, Convolutional Neural Networks and Hidden Markov Models", in *Neural Networks*, 1994.

[73] J. Ma and S. Olsen, "Depth from Zooming", in *Journal of the Optical Society of America*, 7:no.4:1883-1890, 1990.

[74] D. Marr, "Vision", W. H. Freeman and Company, New York, NY, 1982

[75] R.L. Lagendijk, A. Hanjalic, M. Ceccarelli, M. Soletic, and E. Persoon, "Visual Search in a SMASH System", in the proceedings of the *International Conference on Image Processing*, pp. 671-674, 1997.

[76] J. Malik and P. Perona, "Preattentive texture discrimination with early vision mechanisms," in *Journal of Optical Society of America A*, 7 (2), May 1990, pp. 923-932.

[77] B.Manjunath and W. Ma, "Texture Features for Browsing and Retrieval of Image Data," in *IEEE Trans. Pattern Analysis and Machine Intelligence*, vol. 18, no. 8, pp. 837-859, 1996.

[78] B.F.J. Manly, "Multivariate Statistical Methods," *Chapman & Hall*, 1994.

[79] S. Marks and O.J. Dunn, "Discriminant Functions when Covariance Matrices are Unequal", *J. Amer. Statist. Assoc.*, 69, 1974.

[80] G. Smith and I. Burns, "MeasTex", http://www.cssip.elec.uq.edu.au/~guy/meastex/meastex.html.

[81] G.A. Miller, "English Verbs of Motion: A Case Study in Semantics and lexical Memory," in *Coding Process in Human Memory*, Martin and Melton, Eds., Washington, DC: Winston, 1972.

[82] J.R. Miller, J.R. Freemantle, M.J. Belanger, C.D. Elvidge and M.G. Boyer, "Potential for determination of leaf chlorophyll content using AVIRIS", in the proceedings of the *Second Airborne Visible/Infrared Imaging Spectrometer (AVIRIS) Workshop*, pp. 72-77, June 4-8, 1990, Pasadena, Calif. USA.

[83] Y. Awaya, J.R. Miller and J.R. Freemantle, "Background Effects on Reflectance and Derivatives in an Open-Canopy Forest using Airborne Imaging Spectrometer Data", in the proceedings of the *XVII Congress of ISPRS* Aug. 2-14, 1992 Washington, D.C., USA. pp. 836-843.

[84] T. Minka, "An Image Database Browser that Learns from User Interaction," *Masters Thesis, M.I.T. Media Lab Perceptual Computing Group Technical Report No. 365*, 1996.

[85] H.-H. Nagel, "From Image Sequences towards Conceptual Descriptions," *Image and Vision Computing* 6:2, pp. 59-74, 1988.

[86] H. Murase and S.K. Nayar, "Visual Learning and Recognition of 3-D Objects from Appearance", *International Journal of Computer Vision*, 14:5-24, January 1995.

[87] M. Hebert, J. Ponce, T.E. Boult, A. Gross, and D. Forsyth, "Report of the NSF/ARPA Workshop on 3D Object Representation for Computer Vision", Dec. 5-7, 1994.

[88] S. Peleg, J. Naor, R. Hartley, and D. Avnir, "Multiple Resolution Texture Analysis and Classification," *IEEE Trans. Pattern Analysis and Machine Intelligence*, vol. 6, no 4, pp. 518-523, 1984.

[89] A.P. Pentland, "Fractal-based Description of Natural Scenes," in *IEEE Transactions Pattern Analysis and Machine Intelligence*, vol. 6, no 6, pp. 661-674, 1984.

[90] A.P. Pentland, "Shape Information from Shading: A Theory about Human Perception", in *Second International Conference on Computer Vision*, pp 403-413, Computer Society Press, 1988.

[91] M.A. Turk and A.P. Pentland, "Face Recognition Using Eigenfaces", in the proceedings of the conference on *Computer Vision and Pattern Recognition*, pp. 586-591, 1991.

[92] P. Perona, "Deformable Kernels for Early Vision," in *IEEE Transactions on Pattern Analysis and Machine Intelligence*, Vol. 17, pp. 488-499, 1995.

[93] A.P. Pentland, R.W. Picard, and S. Sclaroff, "Photobook: Tools for Content-Based Manipulation of Image Databases," in SPIE proceedings of *Storage and Retrieval for Image and Video Databases II*, Vol. 2, 185, SPIE, Bellingham, Wash., 1994, pp. 34-47.

[94] R.W. Picard, "A Society of Models for Video and Image Libraries," Massachusetts Institute of Technology Media Lab Perceptual Computing Group Technical Report No. 360, 1996.

[95] B. Vijayakumar, D.J. Kriegman, and J. Ponce "Invariant-Based Recognition of Complex Curved 3D Objects from Image Contours", in the proceedings of the *International Conference on Computer Vision*, pp. 508-514, 1995.

[96] M. Flickner, H. Sawhney, W. Niblack, J. Ashley, Q. Huang, B. Dom, M. Gorkani, J. Hafner, D. Lee, D. Petkovic, D. Steele, and P. Yanker, "Query by Image and Video Content: The QBIC System," in *IEEE Computer*, Vol. 28, No. 9, pp. 23-32, September, 1995.

[97] R.J. Qian, M.I. Sezan and K.E. Matthews, "A Robust Real-Time Face Tracking Algorithm", in the proceedings of the *International Conference on Image Processing*, pp. 131-135, 1998.

[98] A.C. Rencher, "Methods of Multivariate Analysis," *Wiley,* 1996.

[99] J.M. Rubin, "Categories of Visual Motion," Ph.D. Thesis, 1977.

[100] R.H. Riffenburgh and C.W.Clunies-Ross, "Linear Discriminant Analysis," *Pacific Science*, 14:251-256, 1960.

[101] D. Saur, Y.-P. Tan, S.R. Kularni, and P.J. Ramadge, "Automated Analysis and Annotation of Basketball Video," in SPIE proceedings of *Storage and Retrieval for Image and Video Databases*, pp. 176-187, 1997.

[102] S. Satoh and T. Kanade, "Name It: Associating of Face and Name in Video", *Carnegie Mellon University Computer Science Department Technical Report CMU CS-95-186*, 1996.

[103] H.S. Sawhney, S. Ayer, and M. Gorkani, "Model-based 2d & 3d Dominant Motion Estimation for Mosaicing and Video Representation," in the proceedings of the *International Conference on Computer Vision*, pages 583-590, 1995.

[104] P.-S. Tsai and M. Shah, "A Fast Linear Shape From Shading Algorithm," in the proceedings of the conference on *Computer Vision and Pattern Recognition*, pp. 734-736, 1992.

[105] B. Shahraray and D. Gibbon, "Automatic Generation of Pictorial Transcripts of Video Programs," in *Multimedia Computing and Networking*, SPIE 2417, pp. 512-528, 1995.

[106] M. Smith and T. Kanade, "Video Skimming for Quick Browsing Based on Audio and Image Characterization," *CMU Computer Science Department Technical Report CMU CS-95-186*, 1995.

[107] D.F. Specht, "Generation of Polynomial Discriminant Functions for Pattern Recognition", in *IEEE Transactions on Electronic Computers*, vol. EC-16, no. 3, pp. 308-319, 1967.

[108] M.J. Swain and D.H. Ballard, "Color Indexing" in *International Journal of Computer Vision*, 7:1, pp. 11-32, 1991.

[109] R. Szeliski, "Video Mosaics for Virtual Environments", IEEE Computer Graphics and Applications, 16(2), pp. 22-30, 1996.

[110] M. Szummer, "Temporal Texture Modeling," Master Thesis, Massachusetts Institute of Technology, Media Lab, 1995.

[111] M. Szummer and R.W. Picard, "Indoor-outdoor image classification," in the proceedings of the IEEE workshop on *Content based Access of Image and Video Databases*, in conjunction with ICCV'98, (Bombay, India), Jan. 1998. http://www-white.media.mit.edu/people/szummer/profile.html

[112] K. Toyama and G.D. Hager, "Incremental Focus of Attention for Robust Visual Tracking," in the proceedings of the *Conference on Vision and Pattern Recognition'*, 1996.

[113] J.K. Tsotsos, J. Mylopoulos, H.D. Covvey, S.W. Zucker, "A Framework for Visual Motion Understanding", in *IEEE Transactions on Pattern Analysis and Machine Intelligence*", Special Issue on Computer Analysis of Time-Varying Imagery", Nov. 1980, p563 - 573.

[114] A. Gupta and R. Jain, "Visual information retrieval", *Comm. Assoc. Comp. Mach.*, 40(5), May 1997

[115] A. Vailaya, A. Jain, and H.J. Zhang, "On Image Classification: City Images vs. Landscapes," in the proceedings of the IEEE workshop on *Content based Access of Image and Video Libraries*, June, 1998.

[116] N. Vasconcelos and A. Lippman, "A Bayesian Framework for Semantic Content Characterization," in the proceedings of the conference on *Computer Vision and Pattern Recognition*, pp. 566-571, 1998.

[117] J.S. Weszka, C.R. Dyer, and A. Rosenfeld, "A Comparative Study of Texture measures for Terrain Classification," *IEEE Transactions on Systems Man and Cybernetics*, vol. 6, no 4, pp. 269-285, 1976.

[118] R.R. Wilcox, *Introduction to Robust Estimation and Hypothesis Testing*, Statistical Modeling and Decision Science Series, Academic Press, 1997.

[119] M. Yeung, B.-L. Yeo, and B. Liu, "Extracting Story Units from Long Programs for Video Browsing and Navigation," in the proceedings of *International Conference on Multimedia Computing and Systems,* 1996.

[120] M. Yeung, and B.-L. Yeo, "Video Visualization for Compact Presentation and Fast Browsing of Pictorial Content," *IEEE Transactions on Cicuits and Systems for Video Technology*, vol. 7, no 5, pp. 771-785, 1996.

[121] D. Yow, B.L. Yeo, M. Yeung, and G. Liu, "Analysis and Presentation of Soccer Highlights from Digital Video," in the proceedings of the *Asian Conference on Computer Vision*, 1995.

[122] H.J. Zhang, S.W. Smoliar, and J.H. Wu, "Content-Based Video Browsing Tools," in SPIE proceedings of *Storage and Retrieval for Image and Video Databases*, pp. 389-398, 1995.

Index